Lecture Notes in Computer Science   8311

Commenced Publication in 1973
Founding and Former Series Editors:
Gerhard Goos, Juris Hartmanis, and Jan van Leeuwen

Editorial Board

David Hutchison
Lancaster University, UK

Takeo Kanade
Carnegie Mellon University, Pittsburgh, PA, USA

Josef Kittler
University of Surrey, Guildford, UK

Jon M. Kleinberg
Cornell University, Ithaca, NY, USA

Alfred Kobsa
University of California, Irvine, CA, USA

T0226215

Friedemann Mattern
ETH Zurich, Switzerland

John C. Mitchell
Stanford University, CA, USA

Moni Naor
Weizmann Institute of Science, Rehovot, Israel

Oscar Nierstrasz
University of Bern, Switzerland

C. Pandu Rangan
Indian Institute of Technology, Madras, India

Bernhard Steffen
TU Dortmund University, Germany

Madhu Sudan
Microsoft Research, Cambridge, MA, USA

Demetri Terzopoulos
University of California, Los Angeles, CA, USA

Doug Tygar
University of California, Berkeley, CA, USA

Gerhard Weikum
Max Planck Institute for Informatics, Saarbruecken, Germany

Anna-Lena Lamprecht

User-Level Workflow Design

A Bioinformatics Perspective

 Springer

Author

Anna-Lena Lamprecht
Universität Potsdam
Institut für Informatik
August-Bebel-Str. 89, 14482 Potsdam, Germany
E-mail: lamprecht@cs.uni-potsdam.de

This monograph is a revised version of the author's doctoral dissertation, which was submitted to Technische Universität Dortmund, Department of Computer Science, Otto-Hahn-Straße 14, 44227 Dortmund, Germany, and which was accepted in October 2012.

ISSN 0302-9743   e-ISSN 1611-3349
ISBN 978-3-642-45388-5   e-ISBN 978-3-642-45389-2
DOI 10.1007/978-3-642-45389-2
Springer Heidelberg New York Dordrecht London

Library of Congress Control Number: 2013955600

CR Subject Classification (1998): D.2, F.3, I.6, H.4, J.3

LNCS Sublibrary: SL 2 – Programming and Software Engineering

© Springer-Verlag Berlin Heidelberg 2013
This work is subject to copyright. All rights are reserved by the Publisher, whether the whole or part of the material is concerned, specifically the rights of translation, reprinting, reuse of illustrations, recitation, broadcasting, reproduction on microfilms or in any other physical way, and transmission or information storage and retrieval, electronic adaptation, computer software, or by similar or dissimilar methodology now known or hereafter developed. Exempted from this legal reservation are brief excerpts in connection with reviews or scholarly analysis or material supplied specifically for the purpose of being entered and executed on a computer system, for exclusive use by the purchaser of the work. Duplication of this publication or parts thereof is permitted only under the provisions of the Copyright Law of the Publisher's location, in its current version, and permission for use must always be obtained from Springer. Permissions for use may be obtained through RightsLink at the Copyright Clearance Center. Violations are liable to prosecution under the respective Copyright Law.
The use of general descriptive names, registered names, trademarks, service marks, etc. in this publication does not imply, even in the absence of a specific statement, that such names are exempt from the relevant protective laws and regulations and therefore free for general use.
While the advice and information in this book are believed to be true and accurate at the date of publication, neither the authors nor the editors nor the publisher can accept any legal responsibility for any errors or omissions that may be made. The publisher makes no warranty, express or implied, with respect to the material contained herein.

Typesetting: Camera-ready by author, data conversion by Scientific Publishing Services, Chennai, India

Printed on acid-free paper

Springer is part of Springer Science+Business Media (www.springer.com)

Foreword

The continuous trend in computer science to lift programming to higher abstraction levels increases scalability and opens programming to a wider public. In particular, service-oriented programming and the support of semantics-based frameworks make application development accessible to users with almost no programming expertise. Indeed, for specialized areas, application programming is reduced to a kind of orchestration discipline where applications arise simply by composing coarse-granular artifacts. User-level workflow design is such an approach. It typically arises in a well-defined environment, and it is often characterized by ad-hoc decisions: "I want to do this just now in this slightly different way." There are numerous methods and tools to support user-level workflow design, most of them tailored to very specific scenarios and with very specific support based on different technologies ranging from scripting approaches, over ontological domain modeling and reasoning in various logics, to techniques and templates used for variability modeling. The jABC developed at the TU Dortmund is an open framework specifically designed to support many of these technologies and approaches. It comprises techniques for constraint-based specification in various logics, model checking, model synthesis, and various forms of model-to-model or model-to-code transformations. In particular, it supports *loose programming*, a method allowing users to write incomplete, (type) incorrect model fragments that are then (if possible) automatically completed to executable models via synthesis technology.

Anna-Lena Lamprecht's thesis establishes *requirement-centric scientific workflow design* as an instance of consequent constraint-driven development. Requirements formulated in terms of user-level constraints are automatically transformed into running applications using temporal logic-based synthesis technology. Anna-Lena Lamprecht illustrates the impact of her approach by applying it to four very different bioinformatics scenarios:

- Phylogenetic analysis, characterized by its huge, publicly available component libraries and an enormous wealth of variants. The model components used here are automatically generated from the service collection EMBOSS using the semantic annotations on top of the EDAM ontology.
- The dedicated GeneFisher-P scenario, which evolved from decomposing the successful GeneFisher tool in order to obtain modeling components that are then manually annotated also on the basis of the EDAM ontology.
- The FiatFlux-P scenario, with its numerous ways of handling large sets of data. It is special, as it requires an annotation beyond the EDAM ontology and some constraints that are difficult to express.
- The microarray data analyses with their typically linear but comparatively long workflows, which directly fit the linear-time temporal synthesis technology and can therefore be fully automatically created by means of jABC's synthesis technology. The microarray scenario requires, however, refinement of the EDAM ontology with semantic annotation in order to arrive at an adequate level of detail.

Conceptually, Anna-Lena Lamprecht's approach adopts the quite simplistic but elegant point of view underlying the eXtreme Model driven Design/Development (XMDD) paradigm, which considers system development as "just" an incremental process of decision taking and concretization, successively introducing more and more constraints, and ultimately leading to a concrete realization. This constraint-oriented approach very naturally subsumes a lot of ideas that have recently been (re)discovered in the area of software variability modeling and product line engineering. There, constraints, be they static/architectural or behavioral, as dominant in the case of workflow design, are used to validate typically manually constructed variants. Anna-Lena Lamprecht's synthesis-based approach goes beyond these checking-based approaches by (additionally) applying synthesis technology to the given constraints in order to achieve correctness by construction. The resulting *constraint-driven* approach provides application experts with the means for a requirement-driven workflow design that is characterized by "playing" with variations of constraints/requirements until satisfactory solutions are synthesized, rather than trying to manually define solutions and then check them for validity.

Technically, constraint-driven modeling and design is supported by the "loose programming" facility of the jABC, which provides application experts with the possibility to graphically compose the intended workflow in their terminology, without caring about technical details like type compatibilities, exchange formats, and connectivity. Even knowledge about the availability of functional components is not required as the user can simply stick to the more symbolic ontological level and leave the service selection to the respective discovery mechanisms. This discovery-based decoupling automatically leads to what is called "open world" assumption in variability modeling, as the user-level constraints seamlessly adapt to correspondingly semantically annotated, evolving service libraries.

This declarative top-down approach to workflow modeling overcomes a number of serious conceptual problems; however, at the price of a very high algorithmic complexity, introduced by a new kind of "filter-based" thinking when searching for adequate workflows. In fact, rather than speaking of workflow modeling, one could in this case perhaps better speak of requirement-driven *workflow mining or workflow discovery*: Users simply provide constraints and obtain the corresponding sets of solving workflows. This may range from the empty set, in case the constraints are not solvable within the given repository, to infinite sets, e.g., in case the constraints allow for unbounded iteration. Anna-Lena Lamprecht carefully investigates the power of this approach, its limitations, pragmatics/heuristics, and the way it may be taken up by non-IT people along a number of case studies and the feedback from life science scholars to her interdisciplinary lectures, which involve intense experimental student work.

There is no doubt that Anna-Lena Lamprecht paves the way for a novel and very powerful way of scientific workflow modeling. Her in-depth investigation of realistic application scenarios based on impressive standard libraries, comprising a realistic level of detail, effectively illustrates what is possible now, where the current limitations are, and what should be done to overcome them. In addition, her careful comparison with more traditional approaches to workflow modeling, which fall far behind when it comes to truly user-oriented workflow modeling, clearly indicates the potential impact of constraint-driven workflow modeling, which may well turn out to be the approach of choice for the future.

Bernhard Steffen

Abstract

Just as driving a car needs no engineer, steering a computer should need no programmer and the development of user-specific software should be in the hands of the user. Service-oriented and model-based approaches have become the methods of choice for user-centric development of variant-rich workflows in many application domains. Formal methods can be integrated to further support the workflow development process at different levels. Particularly effective with regard to user-level workflow design are constraint-based methods, where the key role of constraints is to capture intents about the developed applications in the user's specific domain language.

This book follows the *loose programming paradigm*, a novel approach to user-level workflow design, which makes essential use of *constraint-driven workflow synthesis*: Constraints provide high-level, declarative descriptions of individual components and entire workflows. Process synthesis techniques are then used to automatically translate the high-level specifications into concrete workflows that conform to the constraints by design. Loose programming is moreover characterized by its unique holistic perspective on workflow development: being fully integrated into a mature process development framework, it profits seamlessly from the availability of various already established features and methods.

In this book, the applicability of this framework is evaluated with a particular focus on the bioinformatics application domain. For this purpose, the first reference implementation of the loose programming paradigm is applied to a series of real-life bioinformatics workflow scenarios, whose different characteristics allow for a detailed evaluation of the features, capabilities, and limitations of the approach. The applications show that the proposed approach to constraint-driven design of variant-rich workflows enables the user to effectively create and manage software processes in his specific domain language and frees him from dealing with the technicalities of the individual services and their composition. Naturally, the quality of the synthesis solutions crucially depends on the provided domain model and on the applied synthesis strategy and constraints.

A systematic evaluation of the constraint-driven workflow design approach based on key figures that can be measured during the synthesis process analyzes the impact of the domain model and constraints further. On the one hand, it shows how constraints can directly influence the synthesis results by guiding the search to the actually intended solutions. On the other hand, it shows how constraints can also have a positive effect on the synthesis performance by decreasing the size of the search space considerably. This is particularly relevant since state explosion, that is, the combinatorial blow-up of the state space, is an inherent issue of the synthesis method. Although state explosion cannot be entirely eliminated, it can be tamed effectively by appropriate domain design and constraint application.

Accommodating the experiences gained by working on the application scenarios and their evaluation, finally a set of *loose programming pragmatics* is formulated that provide general guidelines for adequate domain modeling and successful synthesis application. With regard to domain modeling, they emphasize the importance of adequate granularity of service and data types, precise domain-specific vocabulary and clear semantic service interface descriptions in the domain models. With regard to the actual workflow design, they advocate an incremental exploration of the solution space, until an adequate set of solutions is obtained.

In total, the systematic evaluation of the loose programming approach for applicability in the bioinformatics domain steps into a new line of research on user-level workflow design. It has great potential to be continued in different directions, ranging from methodical enhancements of the underlying framework to large-scale, systematic exploration of further application domains.

Acknowledgments

First of all, I am very grateful to my "academic parents" Prof. Dr. Bernhard Steffen and Prof. Dr.-Ing. Tiziana Margaria for their continuous, constructive and active support of my work. Furthermore I am grateful to Prof. Dr. Falk Schreiber and Dr. Mark Wilkinson for helpful discussion and for acting as referees, and to Prof. Dr. Jakob Rehof and Dr. Oliver Rüthing for joining the examination board. Many thanks to all my colleagues of the Chair for Programming Systems ("LS 5"), TU Dortmund, for the great atmosphere in the group, and hence good times in Dortmund. Special thanks go to Stefan Naujokat for being a fantastic office mate and for fruitful joint project work. Many thanks also to Prof. Dr. Lars Blank and Dr. Birgitta Ebert for the good cooperation in the FiatFlux-P project and beyond, and to Prof. Dr. Ina Schaefer for the recent joint work in the context of software product lines and variability management, which raised my awareness of the variability aspects of the methods. Last but not least, I am very grateful to all those who spent some of their time for proofreading (parts of) this thesis and provided valuable feedback, namely, Dr. Birgitta Ebert, Stefan Naujokat, Johannes Neubauer, Sebastian Schäfer, and Julia Rehder.

Contents

List of Figures

List of Tables

Part I

Framework

1

Introduction

This book addresses the challenge of user-level workflow design with a particular focus on the bioinformatics application domain. Towards this aim, it follows a novel, constraint-driven approach to domain-specific design and management of variant-rich workflows and systematically evaluates its features based on a selection of bioinformatics application scenarios. This introductory chapter motivates the work (Section 1.1) and outlines the results and structure of this book (Sections 1.2 and 1.3, respectively).

1.1 Motivation

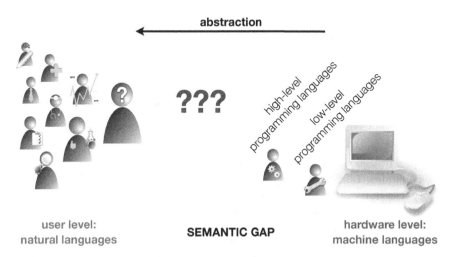

Fig. 1.1 Semantic gap in software development

Just as *driving a car needs no engineer*, steering a computer should need no programmer and the development of user-specific software should be in the hands of the user [208]. Clearly, there is an inherent *semantic gap* between the natural languages that are used for software description at the user level and the machine languages that are required for software implementation at the hardware level. *Abstraction* from implementation details is key for minimizing this gap and eventually lifting software development to an intuitive, user-accessible level.

As depicted in Figure 1.1 in simplified terms, the various existing low-level and high-level programming languages already provide different levels of abstraction from the machine level. They are, however, only sufficiently convenient for programmers, but not for average users, so that still a considerable semantic gap remains that needs to be bridged. Accordingly, research on *end-user development* (cf., e.g., [188, 251, 74, 61]) is concerned with software development methods, techniques and tools that explicitly put the user in the position to develop application-specific software on his own, flexibly according to his needs. In essence, an adequate user-level software development environment has to be:

- *simple enough* to be accessible by the user, and
- *generic enough* to cope with the variability of his applications.

A technology is regarded simple if it is easy to learn for its users and if little programming and IT skills are required for its application. Generality refers to the capability of a technology to be used for different application areas.

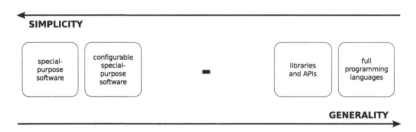

Fig. 1.2 Trade-off between simplicity and generality in software environments

Apparently, however, there is an inherent tradeoff between the simplicity and generality of a technology (cf., e.g., [224, Chapter 18] and [356]), as Figure 1.2 illustrates: Special-purpose software tools are easy to use, but provide only a fixed, limited range of functionality that is hard to extend by the user. In contrast, full programming languages provide a maximum of generality, but demand proper technical or programming skills for fruitful application. Configurable software offers a little more generality, and software libraries and APIs simplify programming to some extent, but there is still a gap between these kinds of software environments that needs to be bridged in order to adequately balance simplicity and generality. Finding the level of balance between simplicity and generality that is adequate for the envisaged application(s) is indeed the central challenge with respect to enabling true user-level software development.

1.1.1 The Workflow Approach to User-Level Software Development

Manifold approaches to end-user software development have been proposed (cf., e.g., [82, 154] for surveys), ranging from, for instance, simplified scripting languages over domain-specific modeling [274, 151] and domain-specific languages [96] to methods for programming by example [287, 187]. Especially in the context of managing variant-rich user-level procedures, many application domains have adopted the notion of *processes* or *workflows* [183, 327].

1. Select picture.

2. (Rotate.)

3. Improve.

4. Add text.

5. Add border.

6. Send to recipient.

Fig. 1.3 Simple workflow example: creation of custom birthday cards

As a simple example, consider a workflow for the creation of custom birthday cards as depicted in Figure 1.3. It consists of six steps: picture selection, rotation (if necessary), quality improvement, adding a personal birthday message, adding a decorative border and finally sending the custom card to the recipient. This workflow can of course be carried out manually by performing the individual steps via the graphical user interface of an image processing tool. However, if the workflow is to be carried out more often, for instance because the CEO of a company wants to send a custom birthday card to every employee at the day of his birthday, then its automation is clearly advantageous. Nevertheless, a fixed single-purpose tool for the procedure would also be inadequate, as workflow variations are likely to be desired.

For instance, custom cards may be created also for other occasions (such as marriage, birth of children, particular achievements, but also as get-well cards or condolence letters), which then require a different selection of pictures and a different message content. The structure of the workflow itself may vary, too: The initial picture may be selected from different sources (such as the local file system, a camera, a database, or a web site), the basic processing steps (rotation, improvement, calibration, resizing etc.) that should be applied depend on the properties of the concrete picture, other kinds of text and decorations may be used (such as more fancy text and border styles, or additional artistic effects), and sending to the recipient can be done in different ways (for instance by e-mail or as printout via inhouse mail). Given these characteristics, it seems in fact natural to regard the example from the workflow perspective.

Technically, a workflow can be defined as "a network of tasks with rules that determine the (partial) order in which the tasks should be performed" [327, Chapter 1]. Consequently, an adequate user-level workflow management system has to provide a user-accessible mechanism for handling both tasks and rules. In fact, service-oriented, model-based approaches, which make use of abstract, intuitive, and usually graphical representations of these tasks and rules, have become the prevalent methods of choice for workflow development across all application domains.

This is commonly known for the development of business processes and scientific workflows, where service-oriented and model-based approaches have become well established in the last years (cf., e.g., [62, 317]). However, also in entirely different application domains, where one would typically expect classical programming styles to be clearly dominant, they are increasingly being used. Just to name a few examples: functionality for graphical scripting is part of the new graphics engine of the "Unreal 4" video game [19], the ConnectIT [34] project provides an intuitive graphical framework for the modeling of ConnectFour game strategies, and the Yahoo! Pipes [12] web application allows for the model-based development of web 2.0 data mashups via a graphical user interface.

Service-Orientation

With the rise of the world-wide web and the spreading of distributed systems, service-orientation [24] with its strong emphasis on proper modularization has established itself as a popular paradigm for IT infrastructures. Its different flavors cover service-oriented architectures (cf., e.g., [31, 323]) in general as well as particular concepts for application development, such as service-oriented software engineering [311], service-oriented computing [133] and service-oriented software design [212].

According to [206], a service is "any nugget of functionality that is directly executable and can be published for use in complex applications". More technically, a service is any (local or remote) resource that is programmatically accessible via a defined interface. In particular, Web Services, as defined by the World Wide Web Consortium (W3C) [49], have become the quasi-standard remote access mechanism in the last years. Nevertheless, several other technologies, such as the Object Management Group's CORBA [241] or the remote procedure call (RPC) facilities [237] included in many programming language APIs, provide equivalent technical support for service-orientation. Finally, service is always "in the eyes of the beholder" [197], that is, the conception as well as the technical realization of services depends on the characteristics and requirements of the specific application scenario.

For the image processing workflow example of Figure 1.3, for instance, the ImageMagick [310, 10] command line tools can be used as basic services. Its flexibly parameterizable `compare`, `composite`, `convert`, `identify`, `mogrify` and `montage` tools provide a comprehensive collection of ready-to-use image processing functionality that, in contrast to common image processing software, runs autonomously without a graphical user interface. Technically, these tools can either be executed directly via the command line interface, or encapsulated into some wrapper script or the like that facilitates integration with other software environments.

Workflow applications regard services as first-class citizens that provide clearly defined, user-accessible units of functionality [197]. This service-orientation clearly facilitates the domain-specificity of the workflow applications, and furthermore inherently supports virtualization and loose coupling of the workflow building blocks [206].

Model-Based Workflow Development

In model-based workflow development frameworks, users can easily build and handle workflows without knowledge of details of the underlying technologies (cf., e.g., [208]). The level of abstraction that is necessary in this respect is essentially achieved by:

- Abstraction from technical details of the single services through virtualization concepts that enable the integration of heterogeneous resources into a homogeneous environment.
- Abstraction from syntactical details of a particular (programming) language through the application of a graphical and thus more intuitive formalism.

The first kind of abstraction goes in line with the service-oriented nature of workflows, and applications profit from the existing (web) service infrastructure of many application domains. Accordingly the loose coupling of services and thus a *lightweight* workflow composition [205] becomes possible. The second kind of abstraction allows the user to focus on the workflow aspects of the application (such as the execution order of services, or transition conditions). The common graphical workflow models are intuitive programmatic realizations of the workflows, directly providing comprehensive user-level documentation of the experimental procedures. They also enable workflow systems to support the adding, removing, or exchanging of services graphically, as well as changes to the control or data flow. In such environments, the agility in the development and the quality and maintainability of the workflow applications are considerably higher than with conventional implementation.

Figure 1.4 shows an exemplary graphical workflow model for the example workflow depicted in Figure 1.3 that makes use of ImageMagick services as

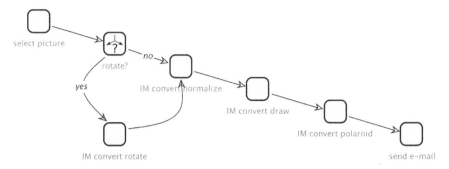

Fig. 1.4 Exemplary image processing workflow model

described above: After selecting the picture for the birthday card, it is decided if it needs to be rotated. If so, a `rotate` service (based on a corresponding function of the ImageMagick tool `convert`) is executed. If not, the workflow directly proceeds to the next step, where `normalize` performs a transformation of the image so that is spans the full range of colors. Then, the custom text is added (`draw`) and the picture set to a style that simulates a Polaroid picture (`polaroid`), before the resulting birthday card is finally sent to the recipient via e-mail.

Note that in contrast to frequently made reservations, the overhead that is introduced by the model-driven development process does not compromise the runtime performance of the workflows (cf., e.g., [175] or [142, Section 5.3.2]): Many workflows orchestrate remote services, that is, the major part of the involved computations is carried out transparently by a remote system, and the workflow itself is mainly concerned with the interfaces of the remote services and the "glue code" between them. Only a small part of the computations takes place on the local machine, and the overall workflow performance is hardly influenced by the local workflow execution performance. In fact, also when the workflow execution is not distributed over a network and no network delays impact the execution times, the majority of the computations is still carried out within the individual services, which can not be influenced by the workflow environment that simply calls them.

1.1.2   Challenges in Scientific Workflow Development

Dealing with the bioinformatics application domain, this book focuses on scientific workflows (cf., e.g., [37, 106, 355, 193, 258]), that is, combinations of activities and computations in order to solve scientific problems. Scientific workflows typically deal with complex, computationally intensive data analysis procedures that are carried out in order to confirm or invalidate a scientific hypothesis [193]. This is in contrast to, for instance, business processes implement real-world procedures concerned with the handling of documents and other business objects across companies and institutions (cf., e.g., [62]). As such, business processes typically involve more communication than computation [106]. The birthday card creation workflow example from the previous section is in this sense hybrid: the actual image processing portions are principally similar to the sequences of computation that are characteristic for scientific workflows, while the final e-mail sending, for instance, is rather a typical business process step. Generally, the following specific characteristics of scientific workflows can be identified [166, 37]:

1. **Rapidly changing workflows**
   Experimental setups in research projects change frequently due to the influence of, for example, new research results, technological progress, or uncertainty in experiments. Accordingly, also the corresponding workflows have to be adapted quite often, for instance by re-ordering sequences of analysis steps, or by exchanging, excluding or including parts of the workflow.

2. **Complex workflows**
   On the one hand, scientific workflows are usually strongly structured. That is, the single steps within the workflow are usually clearly defined and separated from each other. In fact, many popular analysis tools and databases are provided as remote services and hence a plethora of self-contained and easy-to-use (although distributed) resources is available for workflow composition. On the other hand, the entire analysis workflow is usually more complex, for instance comprising several branching conditions and synchronization constraints. What is more, user interaction or a specific interplay of manual and automated lab tasks is often required in the course of the overall experimental workflow. Hence many experiments cannot be carried out fully automatic, but may be supported by semi-automatic workflows.

3. **Specific data types**
   Scientific workflows typically require the use of domain-specific, often complex data structures for adequately representing the various real-world entities they are dealing with. While biological sequences (such as DNA) can easily be represented as strings (though possibly very long ones), the representation of, for instance, spatial structures or measurement results from specific lab devices requires more sophisticated data structures.

4. **High-throughput experimentation**
   Scientific workflows can involve different kinds and large amounts of data: experiments may work on large numbers of data sets, on very large single data items, or on a large variety of data that is derived from original data during the experiment itself. Also the workflows themselves are often subject to high-throughput execution, for instance when the same analysis is executed for a large number of different input data sets, such as on the experimental data that is produced in the labs day by day. Especially, (parts of) workflows that do not require human interaction can

be completely automated and seamlessly integrated into the overall analysis process, which leads to much faster and significantly less error-prone execution of standard data processing tasks. Furthermore, a large number of repeated runs of the same experiment may reveal information about possible experimental errors.

Scientific workflow systems (cf., e.g., [27, 106, 355] for surveys) support and automate the execution of error-prone, repetitive tasks such as data access, transformation, and analysis. In contrast to manual execution of computational experiments (i.e., manually invoking the single steps one after another), creating and running workflows from services increases the speed and reliability of the experiments:

- Workflows accelerate analysis processes significantly. The difference between manual and automatic execution time increases with the complexity of the workflow and with the amount of data to which it is applied. As manual analyses require the full attention of an (experienced) human user, they are furthermore expensive in the sense that they can easily consume a considerable amount of man power. For instance, assume a workflow that needs 5 minutes to perform an analysis which requires 20 minutes when carried out manually. When applied to 100 data sets, it runs for 8:20 h, while the human user would be occupied for 33:20 h, which corresponds to almost a man-week of work. What is more, the automatic analysis workflows run autonomously in the background, possibly also over night, so that the researcher can deliberately focus on other tasks in the meantime.
- Workflows achieve a consistent analysis process. By applying the same parameters to each data set, they directly produce comparable and reproducible results. Such consistency can not be guaranteed when the analysis is carried out repeatedly by a human user, who naturally gets tired and inattentive when performing the same steps again and again. When the analyses are carried out by different people, the situation gets even worse, as achieving consistent behavior of different users is even more difficult to achieve.

Focusing on their actual technical realization, [193] describes the design and development of scientific workflows in terms of a five-phase life cycle (cf. Figure 1.5): Starting point is the formulation of a scientific hypothesis that has to be tested or specific experimental goals that have to be reached. In the subsequent workflow design phase the corresponding workflow is shaped. Services and data must be put together to form a workflow that addresses the identified research problem. It is also possible that (parts of) existing workflows can be re-used or adapted to meet the needs of the new workflow. The workflow preparation phase is then concerned with the more technical preparations (like, e.g., specific parameter settings or data bindings) which are the prerequisites for workflow enactment in the execution phase. Finally there is a so-called post-execution analysis phase, meaning the inspection and

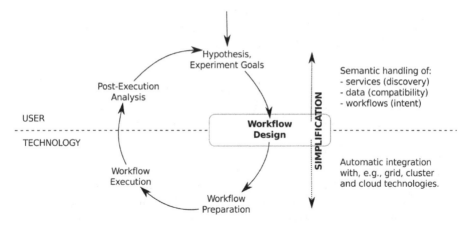

Fig. 1.5 Scientific workflow life cycle (following and extending [193])

interpretation of workflow results. Based on these analyses researchers decide if they have to adapt the experimental setup and to go through the workflow cycle again in order to improve their results. Workflow management systems are typically primarily concerned with the workflow design, preparation, and execution phases.

As Figure 1.5 also shows, the initial hypothesis formulation and the analysis of the results take place at the user level, the workflow preparation and execution phases are located at the technology level, while the workflow design phase is concerned with both user-level and technology-level aspects. Accordingly, simplification of the workflow design phase can be addressed in two complementary directions:

- "upwards", that is, in terms of leveraging workflow design to an even higher and more conceptual level, thus bringing it closer towards the user. This goes in line with the (visions for) workflow systems supporting semantic service composition as described in, for instance, [66, 192, 256, 122, 121, 257, 120, 110, 68, 258].
- "downwards", that is, with respect to achieving a better integration with the underlying technologies, such as grid, cluster and cloud infrastructures, based on the argument that the processing of large amounts of data is crucial (cf., e.g., [27, 355]).

Note that these complementary directions of workflow development do not necessarily have to contradict each other, but can rather complement each other due to a clear separation of concerns: instead of having a single workflow model trying to cover both user and execution logic, it should finally be possible to have an (abstract) model in terms of the user's domain language, which can be automatically translated into highly specialized execution models.

This book is concerned with the simplification of workflow design in the upwards direction, that is, with bringing workflow design closer to the user. The central challenge that has to be addressed in this respect is how to "teach" (domain-independent) workflow management systems to "speak the language of the user". In fact, "many scientists would prefer to focus on their scientific research and not on issues related to the software and platforms required to perform it" [193]. However, the level of abstraction that is common to the present systems still requires quite some technical knowledge about IT and programming in general and the services in particular. Mark Wilkinson's statement that "Many biologists become bioinformaticians out of necessity, not because they like computer science." reflects that the current user group consists of IT-affine application experts and computer scientists rather than of the end-users themselves. Enabling the latter to effectively work with analysis workflows on their own means to leverage the workflow design to an even higher, more conceptual level which truly abstracts from the underlying technologies. That is, application-specific terminology rather than computer science vocabulary should be used to annotate data types and services as described in the previous sections, so that users are able to work with a world-wide distributed collection of tools and data using their own domain language.

As indicated in Figure 1.5 (right) and detailed in the following, this challenge can be addressed by working towards the semantic handling of services, data and workflows. Semantic handling of services provides the basis for user-level workflow design by facilitating the *discovery* of services that are adequate for a particular situation. Semantic handling of data goes further and considers the relationship between services in order to reason about their *compatibility*. The "supreme discipline", however, is their combination with the user's *intents*, expressed via constraints in terms of his domain language, to achieve semantic handling of entire workflows. This allows for assessing the adequacy and correctness of existing workflows and, even more, enables the automatic creation of new workflows that are correct and adequate by design.

All these approaches have in common that they depend on the availability of proper semantic characterizations of the resources in the application domain. That is, a *domain model* [243] is required that provides a formal conceptual model of the application domain. Typically, a domain model comprises a domain-specific vocabulary (usually in the form of an ontology), resources, resource descriptions in terms of the domain vocabulary, a definition of the (possible) relationships between the resources and any additional constraints that characterize the application domain further. In the context of domain modeling for workflow applications, the resources are simply the services and data of the application domain, and the concept of how to relate resources is provided by the workflow management environment, which already determines how services and data can be assembled into workflows. Accordingly, domain modeling in this work focuses on service integration, domain-specific vocabularies, resource descriptions and constraints.

In total, the proclaimed semantic framework constitutes a kind of search engine specialized on services and workflows: Like web search engines such as Google Search [9] help to find single resources in the web, the semantic framework helps to discover *individual resources* in the domain model. The distinguishing characteristic of the semantic framework is, however, that it is designed for finding entire *resource combinations* based on the provided domain model. This is indeed a crucial feature with regard to user-level workflow design, as it supports the automatic identification of service and data combinations that are useful in a particular workflow scenario. The search space for Google is constituted by the vast amounts of content in the World Wide Web, which the search engine manages by indexing and ranking [54]. In contrast, the domain models for workflow applications are as such comparatively small, but the search space that is constituted by the possible resource combinations can easily grow extremely large, which makes the automatic search for workflows inherently complex. And clearly, the obtained results can only be as good as what is provided by the domain model. However, also Google's search engine has not changed the world because it added content – which it did not – but because it has made content easily available to everybody. Similarly, the workflow synthesis method itself does not introduce new resources, but aims at making the workflow potential that is inherent in existing service libraries easily accessible to anyone who needs to design workflows, but especially to non-technical users like biologists and other natural scientists, business process experts and project planners.

Semantic Handling of Services: Discovery

Finding *available* services in a network is a technical challenge that has lead to the development of a number of service discovery techniques, such as the UDDI (Universal Description, Discovery and Integration) registry for web services [20] or the WS-Discovery (Web Services Dynamic Discovery) specification [240]. Finding *appropriate* services within the set of available services is a challenge which goes beyond this standard discovery of services and service interfaces and which has to take into account the semantics of the individual services.

In fact, effective workflow modeling depends on the knowledge of appropriate services for the involved steps. However, it is impossible for a human to overview the plethora of available services manually: The *BioCatalogue* [112], for instance, which has become the central registry for life science web services, contains about 2300 services from around 160 providers at the time of this writing, and is continuously growing. Given such service collections, mechanisms are needed that do not only detect the principally available services, but also help to identify those that are useful in the particular setting. Accordingly, workflow management systems with an intended user group of semi-computer-savvy application experts should provide mechanisms that support these users to find services for their needs, ideally by reasoning about the services in terms of the user's domain language.

Therefore, as pointed out in [127], components should not only be defined via technical, machine-readable low-level descriptions of their program interfaces (the "how"), but also by "the higher-level semantics of the underlying tools, allowing each user to specify his or her requirements in terms of their own, problem-specific terminology" (the "what"). The common way of providing problem-specific or domain-specific terminology is the definition of a controlled vocabulary by means of an ontology, and to use the ontology as classification scheme for the objects of the application domain. In the case of services in workflow management systems, an ontology would provide a service classification scheme, into which all available services are sorted.

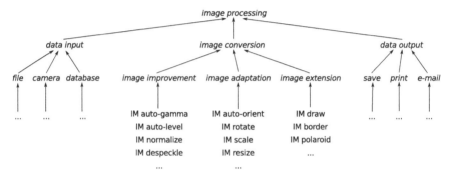

Fig. 1.6 Exemplary semantic service classification

Revisiting the image processing workflow example from Section 1.1.1, a simple hierarchical service classification scheme may for instance look like that depicted in Figure 1.6: the various available *image processing* services are classified further into *data input*, *image conversion* and *data output* services. While *data input* comprises different services for reading pictures from a *file*, *camera* or *database*, the *data output* group contains different services for *saving*, *printing* and *e-mailing* images. The *image conversion* services are then further distinguished into the categories *image improvement*, *image adaptation* and *image extension*, into which then, for instance, the different ImageMagick services (cf. Section 1.1.1) can be sorted. This way, both human users and software applications can easily identify services for a particular purpose by using the terminology of the domain-specific service vocabulary.

A thorough semantic handling of services would at the same time address the issue of service transparency: For instance, the provider-oriented classification of services that is present in many systems leads to the situation that the principally same functionality occurs multiple times in the service list, forcing the workflow developer to decide for one specific server. This may be necessary and useful in some cases, but "it contradicts the idea of preventing workflow developers from having to care how a processing step is performed – a paradigm also promoted by the vision of grid computing, where computing resources are supposed to be transparently allocated from wherever capacity is available" [128]. That is, the ideal workflow system should be capable of transparently choosing one service out of a group of services that provide equivalent functionality, without bothering the user with the technical differences.

Semantic Handling of Data: Compatibility

Numerous data formats have been developed by the scientific communities, reflecting various applications and technical requirements. Their use is continued also when the tools are provided as remote services, meaning that workflows often have to deal with heterogeneous and incompatible data formats. In fact, the heterogeneous and incompatible data formats that are in use constitute one of the main obstacles to service composition and tool interoperation [127]. For instance, there are around 20 common formats for biological sequences alone, and, even more complicated, many available tools and databases use tool-specific ASCII or binary formats rather than one of the more or less common formats. What is more, in the technical terms of the service interfaces, the textual formats are too often only classified as "strings", which is neither apt for reasoning about type compatibility nor does it help users to work with them. Accordingly, workflow systems have to address how to deal with the numerous different data types in a more satisfying way.

In principle, there are two possibilities of how to improve the handling of heterogeneous and incompatible data formats:

1. *Standardization*, that is, introduction of a homogeneous system of more specific data formats. This approach has been taken by several standardization efforts in all application domains. However, a homogeneous standard technology that incorporates all data types is hard to achieve. And even if standards for parts of the data type "jungle" are established, it is impossible to change all the already existing software accordingly in order to thoroughly replace all the historically grown formats.
2. *Automatic adaptation* by adding comprehensive annotations in terms of semantic metadata to the existing data types [301], and using small services that simply perform conversions from one data format into another (so-called "shim services") for achieving compatibility. This approach is indeed more pragmatic than solely striving for standardization, as the annotation is less invasive and can be applied to any resource at any time.

   This way, if standards exist, they can (and should) still be used, but their combination with non-standard data types is also managed.

In analogy to the domain-specific service classifications that have been outlined in the previous section, a detailed semantic description of the data types or data formats that are used by the available services would clearly leverage the usefulness of workflow management systems. Just like strong data typing in programming languages allows for the detection of erroneous function calls, the thorough semantic characterization of service interfaces can support the construction of type-correct, executable workflows [301, 128].

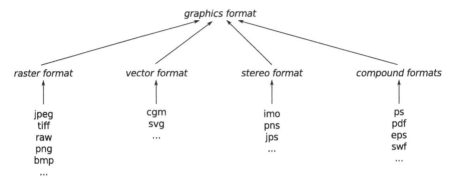

Fig. 1.7 Exemplary semantic type classification

Figure 1.7 shows an exemplary classification scheme for graphics format types, which can be distinguished into *raster formats* (i.e. bitmaps such as jpg, png and bmp), *vector formats* (such as cgm and svg), *stereo formats* (such as imo and pns), and *compound formats* that can contain both pixel and vector data (such as ps and pdf). Based on such type classifications, it is possible to assess the compatibility/incompatibility of services, given that they are properly annotated in terms of the domain-specific type vocabulary.

Fig. 1.8 Semantics-based compatibility assessment

As Figure 1.8 illustrates, this does not only allow to compare data types directly via concrete instances (e.g., jpg is compatible with jpg, but not with svg), but furthermore with awareness of their semantic categorizations. For example, the ImageMagick services can deal with all kinds of bitmaps and may thus be annotated as accepting *raster format* files as input. A service that loads a picture in raw format from a camera can then be combined with the polaroid ImageMagic service, as raw is a *raster format*. In contrast, a service that loads an `svg` image from a database, for instance, can not (directly) be used in conjunction with the ImageMagick services, as they can not process *vector formats*[1].

Semantic Handling of Workflows: Intent

What is more, speaking the language of the user does not only involve the component level, but also the whole workflow. Handling entire workflows in terms of its semantics has two principal flavors:

1. Semantic information about the application domain can be applied for *monitoring workflow development*, essentially by a kind of domain-specific model checking [70], where workflow design is accompanied by the continuous evaluation of constraints that express domain-specific knowledge. Similar to a spell checker in a text processing program, which hints at the violation of orthographic rules, perhaps already indicating a proposal for correction, the model checker notifies the workflow designer when the defined constraints are violated. This provides truly helpful workflow development guidance for the user, beyond mere validation of the workflow model in terms of syntax checks.

2. Semantic information can be used for *semantics-based, automatic workflow composition*: Rather than defining a workflow that already constitutes the solution to a research question, users should only need to express the research question itself using domain-specific terminology, which is then automatically translated into an executable workflow. Thus, not only automation of the workflow would be achieved, but also of its design. Concretely, *declarative* workflow models as proposed, for instance, in [325, 121, 246, 247, 257, 178, 225] can be used to capture the intents about workflows at an user-accessible level. The concretization of the high-level workflow specifications into executable workflow instances can then, for instance, be realized by the application of synthesis-based [252, 163] or planning-based [266, Chapter 10] composition approaches, which work on domain-models that are derived from formally defined service and type semantics and comprehensive (ontological) domain knowledge.

Figure 1.9 illustrates the notion of semantic handling of workflows, again by means of a variant of the simple birthday card creation workflow. Here, the user's intent can simply be expressed as "Take a picture from (his) camera and send it as birthday card to Anna.", as shown in the upper part of the figure. Accordingly, the user may now simply connect an available service that loads a picture from a camera to a service that sends a picture as a birthday card, as shown in the center of the figure. Although this combination of services is

---

[1] Principally, ImageMagick also supports vector formats, but this support is not complete. Hence, the ImageMagick services defined for the example are simply restricted to raster formats to avoid problems.

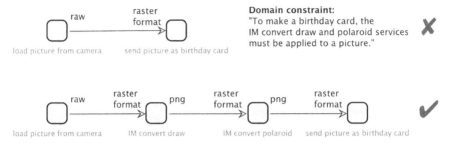

Fig. 1.9 Semantic handling of workflows

type-correct (cf. the compatibility example above), the workflow as a whole is not admissible, as there is a constraint in the domain model which says that "To make a birthday card, the ImageMagick convert `draw` and `polaroid` services must be applied to a picture.".

The four-step workflow at the lower part of the figure is both type-correct and constraint-conform. Instead of being constructed by the user step-by-step until all constraints are fulfilled, it can also be created automatically by exploiting the semantic information provided by the domain model: The incorrect workflow model above can be treated as abstract workflow description that specifies basic information about the intended workflow, in this case two of the services that are to be used. In combination with the interface descriptions of the services in the domain (in terms of their inputs and outputs), the semantic type information and the additional constraints that express further domain knowledge, this incomplete workflow specification can automatically be concretized into a correct and executable workflow.

1.1.3   Technical Consequences

The previous sections identified the primary requirements for a user-level workflow management system as a model-based, service-oriented workflow development environment that is enhanced by semantic methods. This section is concerned with the corresponding concrete technical consequences. In fact, the requirements for scientific workflow systems that are listed in the following are in good alignment with commonly agreed requirements for this kind of system, as described, for instance, in [115, 166, 27, 242, 128, 37, 332] in general and in [256, 120, 258] with particular regard to semantic service environments:

Requirement 1: abstraction from programming details

First of all, abstraction from programming details is essential in order to provide an orchestration level that allows users to cope with the scientific aspects of the experiments rather than with the underlying technology [242].

The researchers that plan the experiments are experts of a certain domain, not skilled programmers. Thus, a helpful workflow management system for scientific research must support intuitive and rapid process design, ideally on a level of abstraction that focuses on defining essential aspects of the workflow (such as the execution order of services, input data, and transition conditions [166]) rather than on technical and syntactical details of a (standard) programming language. This implies a software architecture that separates the engine that deals with the heterogeneous services from the interface that presents a unified world to the end user, ideally by means of a comprehensive and intuitive graphical user interface that makes the system easy to use and facilitates agile workflow development. Graphical workflow representations are furthermore advantageous as they directly provide intuitive documentation of the computational experiment [128].

Requirement 2: powerful workflow model

A powerful workflow model for scientific applications, that is, a model which is fit for implementing complex analysis procedures, has to comprise functionality for control-flow handling, data handling, and hierarchical modeling [128].

- *Requirement 2.1: control-flow handling*
  The ability to define basic aspects like the order of workflow steps and transition conditions between them is of course essential, ideally also synchronization requirements should be expressible [166]. Furthermore, control flow structures like conditional branching and conditional loops are required when building complex workflows. As discussed in [128], pure "for each element"-loops, which allow for iteration over the elements of a list, are in general not sufficient.
- *Requirement 2.2: data handling*
  Typically realizing data analysis processes, scientific workflows rely on input data from which they finally derive (a set of) results, usually also producing a number of intermediate or partial results [166]. Hence, scientific workflow systems have to provide adequate support for defining the flow of data between the services in the workflow. Technically, there are basically two possibilities for passing data between the single services within the workflow: transferring the data via "pipelines" from one service to another (messaging approach) or assigning identifiers to all data items so that the services access them via named variables (shared-memory approach).

  As for the supported data items themselves, it is mandatory to support primitive data types and (one-dimensional) lists [128], but also complex data types and multi-dimensional lists are frequently needed. In order to make it easy for the user to utilize the data handling capabilities of the workflow model, workflow systems should furthermore support basic data processing functionality (such as arithmetic operations, string manipulations and sub-data access methods [128], but also means for data visualization and storage), either via readily provided workflow building blocks or via integration of corresponding functionality into the workflow system itself.

- *Requirement 2.3: hierarchy*
  A hierarchy concept that allows for encapsulation of sub-workflows into reusable components is mandatory for scalability [128]. It should be possible to build new workflows using (entire) existing ones, for instance by executing several workflows consecutively or by applying sub-workflow facilities in order to work with different levels of abstraction [166].

Requirement 3: domain modeling support

As sketched above, domain modeling for scientific workflow applications principally involves the integration of domain-specific services and the semantic enrichment of the domain model by definition of domain-specific vocabularies, semantic resource descriptions and constraints. Both aspects should be supported by a workflow system in order to make it possible to deal with also the most heterogeneous environments in a homogeneous and user-accessible way.

- *Requirement 3.1: service integration*
  Workflow systems should obey to an open world assumption for services [242], that is, allow for the integration of new services. While the set of predefined process building blocks is naturally limited, discovery mechanisms for popular service providers can increase the number of available (web) services dynamically. However, it can not be foreseen in general which kind of local or remote services will be needed. Hence a workflow management system should enable programmers to extend the library of workflow building blocks, for instance by providing special adapters to third-party libraries or arbitrary local and remote services.
- *Requirement 3.2: semantic enrichment*
  Workflow systems should provide the means for defining domain-specific vocabularies for the workflow applications, for instance via taxonomic or ontological classifications of the services and data types in the domain. Semantic annotations of service interfaces in terms of these vocabularies can then provide (both human- and machine-understandable) meta-information about the resources in the domain that can not only help the user to search for the right service for a task, but furthermore makes it possible to automatically find the most suitable tool or select appropriate alternatives in case that a particular service is temporarily not available. In addition, it should be possible to formulate constraints that express further domain-specific knowledge, such as dependencies between particular services or general characteristics of the envisaged applications.

  Just as the service library of a workflow system should be extensible, it should also be possible to extend and customize the semantics-related parts of the domain model according to specific applications. In

particular, the user himself should be given the possibility to bring in his expert knowledge, for instance by classifying services and formulating constraints.

Requirement 4: semantics-based service composition support

Workflow systems should exploit the available semantic information about the services to help the user in finding valid and adequate service combinations. Approaches can principally be distinguished by whether they aim at discovering single suitable services or finding appropriate service sequences:

- *Requirement 4.1: service discovery*
  Service discovery techniques provide means for identifying single services according to some criterion. In the basic case, the criterion can be a search term or an ontological classification. More sophisticated is the discovery of services based on their input and output data types, for instance for identifying the services that match with other services in the workflow.
- *Requirement 4.2: automatic workflow composition*
  (Semi-) automatic workflow composition techniques aim at finding sequences of services according to some higher-level specification. Therefore, it should be possible to express the (global) intents and goals of the workflow at an abstract, semantic level that does not deal with the technical aspects of the concrete workflow realization. Typically, this specification is based on the input and output data types of the intended sequence. Ideally, however, the specification should allow for a fine-tuning of the intents beyond a mere input/output description of the workflow.

  Furthermore, a mechanism is required that translates the abstract workflow description into a concrete workflow realization automatically. This involves the identification of the required services and their composition and configuration in order to form an executable workflow.

Note that automatic workflow composition techniques can often directly be used for service discovery as described above by restricting the search to service sequences of length 1.

Requirement 5: workflow validation and verification

Workflow systems should support the validation and verification of the workflow model with respect to both their static properties and their runtime behavior:

- *Requirement 5.1: static properties*
  Syntax and type checks at the local level as well as model checking techniques that consider the workflow model as a whole may prevent misconfiguration of workflow building blocks at modeling time.

- *Requirement 5.2: runtime behavior*
  Simulation and debugging functionality enables the detection and analysis of workflow modeling errors that become obvious during execution. Ideally the workflow system should also keep track of the workflow executions in a kind of event history, logging which workflows were executed when and with which result. This information can then be used for the analysis of the workflows in terms of, for instance, load factor or execution time statistics, as well as frequently occurring errors or performance issues [166].

Requirement 6: workflow execution

Execution is of course central for scientific workflow applications [242, 128]. Therefore the workflow model should not only represent the underlying process, but rather contain also all information that is necessary for its execution, for instance by using an associated interpreter or a specific execution engine. Such information may be provided at the level of the process building blocks or, globally, at the workflow model level.

Generally, workflow execution can take place within or outside of the workflow environment, and ideally both possibilities should be supported:

- *Requirement 6.1: execution within the workflow environment*
  Execution within the workflow system is convenient when the workflow developer is also the user of the workflow, as he can continue to work in the familiar environment. Furthermore, runtime validation becomes easily possible (see above). Ideally, ad-hoc execution of the workflow model should be carried out by an interpreter component that is provided by the workflow environment.
- *Requirement 6.2: execution outside of the workflow environment*
  Execution outside of the workflow system is required in case that acceptable computation times can only be obtained when the workflow is executed, for instance, on a grid or cluster infrastructure [166]. This requires the deployment of the workflow to, for instance, a workflow engine or the compilation of the workflow model into a stand-alone application, which is ready to be used by others. Therefore code generation and application deployment functionality should be provided by the workflow modeling framework.

This list of requirements is in fact rather general and as such applies to different scientific application domains. More specific requirements, especially those related to supported data formats and styles, are subject to the nature and complexity of the concrete application for which the system is to be used. In fact, many requirements for workflow systems that are identified in the general computer science or business processes context (such as work lists, role assignments, time-based triggers, or deadlines) are considered less relevant for the scientific domain [166].

1.2 Results

In the context described in the previous section, this book focuses on workflow management for the bioinformatics application domain, which is an interesting "playground" for workflow methodologies for different reasons:

- Processing the large amounts of often heterogeneous data that is produced by the life-science laboratories requires computational analysis procedures (in bioinformatics jargon also called "in silico experiments", referring to and contrasting the traditional *in vivo* and *in vitro* experimentation in biology). Rather than having a single piece of software ready for the job, typically combinations of several resources and tools are required for performing the possibly very complex, variant-rich analysis processes.
- The workflow approach to in silico experimentation enjoys great popularity in the community, and accordingly a lot of research has been going on in this field in recent years (cf., e.g., [92, 148]).
- In order to make tools and data easily available to researchers, the community has on the one hand adopted the use of (Web) Services for making tools and data remotely accessible for the public, and on the other hand many resources are open source and can deliberately be used. Thus, both comprehensive data and service repositories are available and facilitate the experimentation with realistic bioinformatics workflow scenarios.
- The bioinformatics community is one of the most advanced with respect to the emerging Semantic Web [40], actively working on providing domain-specific meta-data for all kinds of resources [57]. Hence, the application of semantics-based methods is largely facilitated.

As such, the bioinformatics application domain is particularly suited for considering workflow management methodologies with a focus on variant-rich scientific applications. Owing to the observation that especially with regard to the workflow design phase, future developments must aim at lowering the technical knowledge that is required for successfully combining services into workflows (cf. Section 1.1.2), this book addresses in particular the question how workflow systems can offer better means for finding appropriate services, for dealing with data formats, and for using the technical language of the user rather than computer science terminology for communicating the user's intents about his workflows. Therefore, this comprises two principal parts of work, which are detailed in the following two sections:

1. The conception of a concrete software framework for user-level design of workflows based on semantic techniques.
2. Comprehensive applications of the developed methods in the bioinformatics domain in order to assess their capabilities and limitations.

In total, the systematic evaluation of the loose programming approach for applicability in the bioinformatics domain steps into a new line of research on user-level workflow design. It has great potential to be continued in different directions, ranging from methodical enhancements of the underlying framework to large-scale, systematic exploration of further application domains.

1.2.1   A Framework for User-Level Workflow Design

The conception of the workflow framework presented and used in this book was guided by the ideas that have also been summarized as the *loose programming paradigm* [178]. Loose programming proclaims a form of model-based graphical software development that is designed to enable workflow developers to design their application-specific workflows in an intuitive style. In particular, it aims at making highly heterogeneous services accessible to application experts that need to design and manage complex workflows. After an adequate domain modeling, application experts should ultimately be able to profitably and efficiently work with a world-wide distributed collection of services and data, using their own domain language. Moreover, loose programming enables users to specify their intentions about a workflow in a very sparse way, by just giving intuitive high-level specifications, because there is a mechanism available that automatically translates this request into a running workflow.

In fact, the concept of *loose specification* is key to the loose programming approach: a specific graphical formalism allows developers to express their workflows just by sketching them as kinds of flow graphs without caring about types, precise knowledge about the available process components or the availability of resources. They only have to specify the rough process flow graphically in terms of ontologically defined semantic entities. These loose specifications are then concretized to executable workflows automatically by inserting missing detail. This is achieved by means of a combination of different formal methodologies: *Data-flow* analysis provides information on available and required resources, which is then used by a temporal-logic *synthesis* algorithm [305] to find sequences of services that are suitable to concretize the loose parts. Additionally, *model checking* is used to monitor global process constraints continuously.

In order to address particular open issues of the workflow design phase as identified above, emphasizing the importance of semantics-based workflow development support, the *jABC framework* [306] for model-driven, service-oriented workflow development was extended by functionality for semantics-based, (semi-) automatic workflow composition according to the loose programming paradigm. In contrast to earlier implementations and applications of the synthesis method (such as, e.g., [98, 302, 307, 204]), loose programming assumes a shared memory for data exchange between the services in a workflow. Concretely, it makes use of the control-flow models and workflow execution context as provided by the jABC framework, and applies data-flow analysis techniques to keep track of the available data. As such, workflow design is not restricted to creating mere pipelines of services, but opens a new dimension of workflow synthesis: Data that is created by one service can not only be used by its direct successor, but is available to all subsequently called services, which enables further service compositions. Control-flow structures can only be modeled manually in the current framework, and not be synthesized automatically. They are, however, fully recognized by the data-flow analysis, which thus makes sure that all available data is taken into account when the synthesis concretizes a loose branch at any place in the workflow model.

As a result of these enhancements of the jABC framework, well-established workflow management technologies are provided together with future-oriented, semantic approaches to workflow design within one coherent framework. Its broad range of applications includes complex, long-living applications like the Online Conference Service (OCS) [238], as well as ad-hoc workflows [335] like the exemplary bioinformatics applications, which can be dynamically adapted to changing experimental requirements.

In fact, the implementation of the automatic workflow composition functionality was essentially driven by the requirements identified when working on bioinformatics workflows based on the jABC framework and related technologies. While the initially available, jABC-based comprehensive framework for service-oriented, model-driven design, execution, verification and deployment of bioinformatics workflows has been coined *Bio-jETI* already by [201], the concepts for its extension by functionality for semantics-based, (semi-) automatic workflow composition [174] are the result of the work that has been carried out in the scope of the doctoral dissertation that constituted the basis for this book. Among bioinformatics workflow management systems, Bio-jETI is unique in its application of different flavors of formal methods for workflow design, which is facilitated by the formally defined structure of the workflow models of the underlying jABC framework.

1.2.2 *Applications and Evaluation*

In order to explore the characteristics, capabilities and limits of the loose programming approach to user-level workflow design, the framework described above was applied to different real-life bioinformatics workflow scenarios. Bringing it into application in a particular domain is in the first place a comprehensive *domain modeling* task: The jABC framework and the associated technologies have been developed in a domain-independent fashion, providing general functionality for workflow modeling, such as a graphical editor and mechanisms for persistence, execution, compilation, validation and deployment. In order to use it for a specific application domain, it has to be prepared for use in the target domain by selecting a suitable set of plugins and providing adequate workflow building blocks (services) as well as semantic meta-information about these services in terms of taxonomies or ontologies.

The Bio-jETI incarnation of the jABC framework that is used in the scope of this book is principally formed by its characteristic set of plugins. The domain models then tailor the available workflow building blocks and semantic meta-information further to the concrete envisaged bioinformatics applications, and thus enable the actual domain-specific design of in silico experiments. Accordingly, the technical contributions of this work to the application of the novel workflow design approach to the different in silico experimentation scenarios comprise in particular:

- The semantic enrichment of the domain models, that is, the semantic annotation of service interfaces and semantic classifications of services and data types that facilitate (semi-) automatic, constraint-driven workflow design and largely free workflow designers from dealing with service interfaces and data types. By lowering the required technical knowledge in this fashion, highly heterogeneous services become accessible to researchers who are not specially trained in programming, but who need to design and manage complex bioinformatics analysis workflows.

- The identification and formalization of frequently occurring workflow constraints and domain-specific knowledge beyond the more technical semantic interface descriptions in order to improve automatic workflow composition. Semantic service interface annotations are per se only sufficient for automatically creating generally possible service compositions, hence often (especially when working with large domains) additional knowledge is required for constraining the search to the actually intended workflows. This is specifically supported by the constraint language of the loose programming framework.

In contrast to earlier applications of the synthesis method, such as the LaTeX process synthesis (toy) examples [98, 307] or the Telco applications [304, 297], the bioinformatics workflow scenarios as addressed in this book deal with an existing, historically grown, and in particular less rigidly designed domain that comprises truly distributed services and does impose farther reaching challenges on the workflow design method. Another central difference is the relevance of data in the scientific applications, which demands for a shared memory for flexibly exchanging data between services, and consequently also for the integration of data-flow analysis methods into the synthesis process in order to be able to keep track of the available data throughout the workflow.

The four selected application examples that are discussed in this book cover different thematic areas (phylogenetic analyses, PCR primer design, metabolic flux analysis and microarray data analysis), different software components (commonly known tool collections as well as special-purpose services), different service technologies (standard REST and SOAP Web Services as well as specifically created jETI [202] services), and also workflows of different complexity. In contrast to the more academic case studies carried out for the conceptual study of the application domain, these projects address specific, new application scenarios and provided experiences regarding the practical applicability of the developed methods. Each example has its specific characteristics, and there are considerable differences between the applications:

- **Application scenario 1: phylogenetic analysis workflows**
  The first application scenario is concerned with phylogenetic analyses based on molecular sequences, which have become a standard application domain for illustrating bioinformatics workflow technologies (cf., e.g., [256, 134, 155, 172, 265, 165]). Accordingly, there is a plethora of easy-to-use software tools available for the individual analysis steps, and the workflows of this example could be built exclusively using publicly available services for the computations. This is in contrast to the other scenarios, where specifically designed services were required for certain analysis steps.

  Not least due to the large number of available services, phylogenetic analysis workflows are extremely variant-rich. Hence, they are well suited to demonstrate the agility of workflow design in Bio-jETI and in particular the features and benefits of the constraint-driven, semi-automatic workflow design approach. Furthermore, the annotation with semantic meta-data is particularly advanced for this discipline [179], which is also advantageous for the application of the constraint-driven workflow design methodology.

- **Application scenario 2: GeneFisher-P**
  The GeneFisher-P workflows [177] were developed in order to provide a flexible alternative to the monolithic GeneFisher web application for PCR primer design [109, 124]. The workflow version of the application makes it easy, for example, to change individual steps (e.g. using alternative sequence alignment algorithms), to increase or decrease the amount of required user interaction (e.g. by letting the user assess each intermediate result or by limiting user interaction to the specification of the input data), or to define batch processing workflows that perform primer design for several input sequences autonomously (rather than requiring manual execution for every single input sequence).

  The actual sequence of analysis services that has to be applied for performing primer design depends on the specific type of the input data, which is however not known at application modeling time. The original application and workflow models thus include points of conditional branching, where the user is asked to provide the required data characterizations at runtime, and accordingly an appropriate sequence of services is executed. Using the synthesis functionality of the new framework, the user can generate the services sequence best suited for his concrete input data ad hoc, enabling him to exhaust all possibilities provided by the possibly evolving service libraries.

- **Application scenario 3: FiatFlux-P**
  FiatFlux-P focuses on the automation of metabolic flux analysis. It is designed to work off large sets of data from ^{13}C tracer experiments based on a specialized version of the interactive FiatFlux desktop analysis software

[353]. It integrates well into the entire data analysis process, starting from the raw data sets and spreadsheet experiment documentations from the lab, and producing outputs that can be processed by a specific visualization software.

As there is only a small number of actual analysis services involved, the variants of the FiatFlux-P workflows are mainly imposed by the possible variations in data handling. Nevertheless, these variants are numerous and often run off the actual track, so that for synthesizing FiatFlux-P workflows comparatively many constraints have to be provided to take the relevant dependencies into account and characterize the intended solution(s) appropriately.

- **Application scenario 4: microarray data analysis workflows**
  The microarray data analysis workflows provide a framework for user-level construction of analysis pipelines based on appropriately wrapped and integrated Bioconductor [105] functionality. It helps handling the variability that is inherent in microarray data analysis at an user-accessible level and thus again emphasizes the agility of the model-driven and service-oriented approach to workflow design.

  Remarkably, the workflows in this scenario comprise comparatively many analysis steps, but typically have a simple, in fact mostly linear, structure. Thus, the automatic workflow composition functionality of Bio-jETI, which is based on a linear-time logic synthesis algorithm, can be applied here to generate the complete analysis pipelines automatically.

Although the applications address different scenarios with different characteristics and make use of highly heterogeneous services, they are still realizable homogeneously within the Bio-jETI framework. By using a coherent workflow formalism, it becomes possible to focus on the application-level aspects of the different workflows. In fact, the gained insights about the (technical, service-level) characteristics of the application domain(s) – systematically and adequately captured and formalized – enhanced the knowledge base required for the successful application of semantic techniques to support workflow design.

Furthermore, the different applications provide a proper basis for further research and considerations about bioinformatics workflows in particular and scientific workflows in general. Most importantly, in this work they are used as basis for the first systematic analysis and evaluation of the central synthesis method with regard to performance and the impact of domain model and constraint patterns. To this aim, an evaluation framework has been conceived for analyzing the impact of constraints on results and performance systematically.

On the one hand, the evaluation shows that constraints can directly influence the synthesis results by guiding the search to the actually intended solutions. On the other hand, it shows that constraints can also have a positive effect on the synthesis performance by decreasing the size of the search space considerably. This is particularly relevant since state explosion, that is, the combinatorial blow-up of the state space [324], is an inherent issue of the synthesis method. Although state explosion can not be entirely eliminated, it can be effectively postponed by appropriate domain design and constraint application. Thus, proper semantic domain modeling is obviously crucial. At the same time, adequate domain modeling is clearly difficult, since it has to take into account a plethora of aspects and furthermore depends massively on the characteristics of the concrete application domain.

In this light, the *loose programming pragmatics* formulated in this book summarize the experiences gained by working on the application scenarios and their evaluation and provide useful general guidelines for adequate domain modeling and successful synthesis application. In short, they state that:

1. Services and data types of *adequate granularity* are required as basis for any domain model and workflow application.
2. Semantic domain modeling has to focus on the definition of a *precise domain vocabulary* and the definition of *simple semantic service interface descriptions*.
3. For the actual workflow design, it is advisable to *increase the search depth gradually* and to *specify the constraints incrementally*.

Note that as in all software engineering processes, finally a good amount of experience is required to obtain adequate solutions, and that also the most carefully designed domain model can not be expected to suit all possible application scenarios equally well. A distinguishing feature of the loose programming framework in this regard is that it explicitly encourages the workflow designer to bring in his specific domain knowledge and adapt the domain model according to his particular needs.

1.3 Outline

Concretely, the remainder of this book is structured as follows:

- *Chapter 2* introduces the initially available Bio-jETI framework in detail, before it deals with adding semantic awareness to workflow management, especially concerning the support of workflow design by means of constraint-based workflow development methods. In particular, the chapter explains the extension of Bio-jETI with constraint-driven methods in order to provide domain-specific support for (semi-) automatic workflow design according to the principles of the loose programming paradigm.
- *Chapters 3 – 6* illustrate Bio-jETI's application to the development of bioinformatics workflows by means of four different application scenarios from different thematic areas (phylogenetic analyses, PCR primer design, metabolic flux analysis, microarray data analysis) and with different technical characteristics. For each of the four application scenarios, the respective chapter comprises three major parts:

1. Introduction to the addressed biological questions and the associated analysis methods and software.
2. Presentation of exemplary service libraries and workflows for the application scenario. Note that although this is not shown explicitly, *executable* workflows are developed in all examples.
3. Application of the constraint-driven workflow design methodology to the workflow scenario. Therefore, it is shown how the respective domain models are equipped with adequate semantic annotations and how they are used for (semi-) automatic composition of already known and of new workflows. The developed domain models are (deliberately) not "perfect" (if this is actually possible), but thus they are suitable to illustrate crucial aspects of the domain modeling.

Note that the process of domain modeling and workflow design with Bio-jETI is described most elaborately for the first example application (Chapter 3). In particular, Sections 3.3.3 and 3.3.4 discuss and illustrate in detail an exemplary workflow composition problem and solution refinement strategy, showing the considerable impact that constraints can have on the obtained solutions. This demonstrates how the framework enables the user to "play" with synthesis configurations and constraints in order to explore the solution space interactively and finally arrive at an adequate set of constraints. In order to avoid unnecessary redundancies, the presentation of the other examples is shorter in this regard and restricted to the summarizing description of the finally applied constraints.

- *Chapter 7* gives an analysis and evaluation of the loose programming applications, drawing first conclusions and highlighting distinctive features. The "lessons learned" from the work on the applications are furthermore used to assess the capabilities and limitations of the developed methodology and to formulate pragmatics that provide guidance for its successful application.
- *Chapter 8* discusses the relation of Bio-jETI to other approaches to bioinformatics workflow management. Therefore, it reviews a selection of different workflow systems that have been applied to or designed for the bioinformatics domain and compares them to Bio-jETI. Additionally, it investigates the substantial differences between control-flow-oriented and data-flow-oriented workflow modeling.
- *Chapter 9* concludes this book with a summary and a discussion of remaining challenges and their implications for future work.

The Bio-jETI Framework

Bio-jETI [201, 174] is a comprehensive framework for management of variant-rich bioinformatics workflows that has largely been conceived in the scope of this work. It is based on the jABC framework [306, 229] for model-driven, service-oriented workflow development. Providing comprehensive means for the design, execution and deployment of workflows, already the standard jABC framework covers the common aspects of workflow management.

The distinctive feature of Bio-jETI in contrast to other scientific workflow systems is the incorporation of two different notions of constraint-based methods into the workflow development process, also illustrated by Figure 2.1:

- *Constraint-guarded workflow design* is a development style that is facilitated by the available model-checking functionality. It provides means for reasoning about workflow properties at the user level: workflow design is

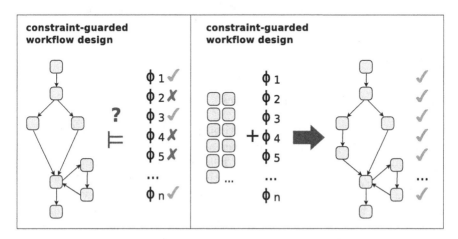

Fig. 2.1 Constraint-based workflow design

accompanied by the continuous evaluation of workflow-level constraints, alerting the workflow designer when constraints are violated.

- *Constraint-driven workflow design* is enabled by the functionality for synthesis-based loose programming by which the framework has been extended in the scope of this work: the application of constraints is furthered, systematically using them for high-level descriptions of individual components and entire workflows, which can then be translated automatically into concrete workflows that conform to the constraints by design. This allows users to describe the intended workflow at an abstract, less technical level by providing simple means for sketching workflow "skeletons" and for expressing workflow constraints in terms of domain-specific vocabularies.

In Section 2.1, this chapter introduces the basic jABC framework and the plugins that are commonly applied when using the framework for model-driven, service-oriented workflow development. Sections 2.2 and 2.3 then focus on the constraint-guarded and constraint-driven workflow design methodology, respectively.

2.1 Model-Driven, Service-Oriented Workflow Development

Bio-jETI builds upon a multi-purpose domain-independent modeling framework, the *Java Application Building Center (jABC)* [306, 229], for the workflow definition and management, and the *Java Electronic Tool Integration* framework *(jETI)* [301, 202] for dealing with the integration and execution of remote services. Both are based on well-established software technology and have been applied successfully to different application domains (cf., e.g. [141, 162, 160, 161, 34, 220, 142]). The following Sections 2.1.1 and 2.1.2 describe the jABC framework and the jETI technology, respectively, in greater detail.

2.1.1   The jABC Framework

Generally speaking, the jABC is a multi-purpose, domain-independent modeling framework. In particular, it has been applied as a comprehensive service engineering environment [206, 197, 210] following the eXtreme Model-Driven Design (XMDD) [208, 209, 157] and Continuous Model-Driven Engineering (CMDE) [210] paradigms. XMDD and CMDE put the (user-level) models into the main focus of software development, technically building upon a combination of extreme programming [39], model-driven design, and process modeling methodologies.

In its simplest form, the jABC is basically a graphical editor for constructing directed, hierarchical graphs by placing nodes (called *Service Independent*

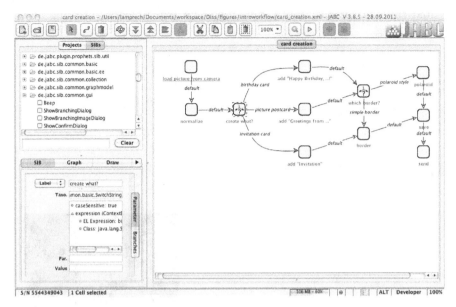

Fig. 2.2 Graphical user interface of the jABC

Building Blocks, or simply *SIBs*) on a canvas and connecting them with labeled edges (called *branches*). Figure 2.2 shows its user interface: the available
SIBs are listed in a browser (upper left), from where they can be dragged
onto the "drawing" area (right), where the model construction takes place.
Different inspectors (lower left) can be used for the detailed configuration of
components and models. With this graphical interface, models can be recon-
figured and extended very easily.

Additional functionality is brought to the framework by means of plugins.
In particular, any kind of interpretation of the model is subject to specific
plugins. As discussed in detail in [229], the basic jABC framework can be
extended by two principal kinds of plugins:

1. *Semantics plugins* provide functionality for specific interpretations of the
   graphical model or its components.
2. *Feature plugins* extend the range of general, domain-independent functionality that is available in the jABC.

In the light of interpreting the graphical models as *Service Logic Graphs*
(*SLGs*), that is, as executable control-flow models composed of services, Bio-
jETI comprises the following semantics plugins:

- The *Tracer* plugin [86] provides an interpreter and execution environment.
  It interprets the models of the jABC as directed graphs that express the
  control-flow of a process. The Tracer allows for the stepwise or complete
  execution of the SLGs, and the use of breakpoints. It provides detailed
  information on the execution history, the objects in the execution context,

and running threads, and is thus useful for experimenting, testing and
debugging.

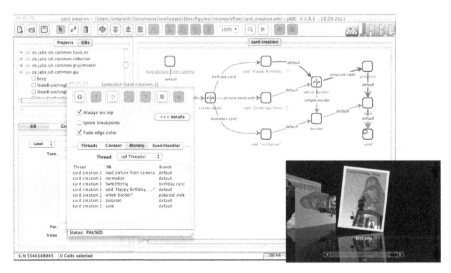

Fig. 2.3 Workflow execution via the Tracer plugin

Figure 2.3 shows the jABC user interface during the execution of a
workflow model with the Tracer: The plugin provides a small "control
panel" (left), which can be used, for instance, to start the complete execution of the workflow, to steer its step-by-step execution, or to stop at
breakpoints. Moreover, it provides basic monitoring facilities, such as an
execution history (as shown in the figure) and an overview of the running
threads, the objects in the execution context and events (not visible in
this figure). The current state of the execution is in addition visualized
in the workflow model: the currently executed SIB is marked by a small
green overlay icon, and the four recently followed branches are colored
green. In the figure, workflow execution has passed the SIBs that execute
the image processing steps for creation of custom birthday cards as described in Section 1.1.1, and the result is now available and ready to be
sent.

In order to use the Tracer for executing an SLG, all its SIBs have to
implement the Tracer interface. This requires the corresponding SIB to
provide an `execute()` method that contains the code that is to be called
during execution, possibly making use of the parameters of the SIB. This
code can be provided completely in the SIB, but also use libraries, or call
remote services. Furthermore, this method has to define a return value
that has to be one of the labels for outgoing branches and represents
the result of the method invocation, that is, for instance, whether it was

successful or not (typically used labels therefore are `default` and `error`, respectively).

- The *Genesys* plugin, a jABC plugin for using the Genesys code generator framework [144, 142], can be used to automatically compile any executable workflow model into a single deployable application that can be run independently of the jABC. The generated application can be, in the easiest case, a standard Java application or Servlet, which can be deployed to any platform for which a Java virtual machine is available (nowadays ranging from mobile phones to supercomputers). The code generators that are used within Bio-jETI also interpret jABC models as control flow graphs, using principally the same mechanisms as the Tracer does.

  Genesys supports two major classes of code generators: *Extruders*, which directly use the execution facilities of the jABC framework (providing features like thread and event handling, and special types of SIBs and SIB parameters), and *Pure Generators*, which compile for the runtime engine provided by a platform of choice (like, e.g., the Java Runtime Environment (JRE)), and where the resulting code is totally independent of the jABC framework.

- *jETI* (Java Electronic Tool Integration) [202] is a platform for easy deployment of file-based command line tools as remote services. A more detailed description of the platform is given in the next section. The jETI plugin is used to accomplish the communication with jETI services.

- With the *LocalChecker*, SIB configurations can be validated during workflow development. The LocalChecker provides means to specify properties, preconditions, and environment conditions at the component level. Examples of local check properties are the correct setting of parameters and the correct and complete connection of a SIB in the surrounding model (e.g., no dangling mandatory exits for a SIB) – cases where erroneous configurations would hamper the correct execution of the control flow.

- The *PROPHETS* plugin [232, 234] enables constraint-driven workflow design according to the *loose programming* paradigm of [178]: rather than requiring workflow designers to model everything explicitly, the workflow model does only have to be specified roughly in terms of ontologically defined "semantic" entities. These loose specifications are then automatically concretized to fully executable concrete workflows by means of a combination of different formal methods. The PROPHETS plugin is introduced and discussed in detail in Section 2.3, where the topic of constraint-driven workflow composition in the jABC is addressed.

Additionally, Bio-jETI makes use of the following feature plugins:

- The *TaxonomyEditor* plugin can be used to define special-purpose classifications of the SIB libraries, called taxonomies. By default, the jABC presents the available SIBs according to their location (as Java classes) in the Java package hierarchy. Taxonomies group services according to specific classification criteria or viewpoints, such as function, origin, technology or input/output behavior. They can be seen as very simple forms of ontologies, namely with a built-in *is-a* relation (subset concepts).

- An important characteristic of the jABC is the formally defined structure of its models, which facilitates the application of different flavors of formal methods. For instance, in addition to the component-level validation that is performed by the LocalChecker, it is also possible to verify global properties of a workflow model. Section 2.2 details how this constraint-guarded workflow design is facilitated by the *GEAR model checking plugin* [35, 36].

- The *SIBCreator* plugin assists in the development of new SIBs. It can be used in a top-down as well as in a bottom-up mode: In the top-down mode, the SIB is specified via the SIBCreator, including information such as name, parameters, branches, and implemented interfaces. From this specification a skeletal SIB class is generated that then has to be equipped with execution code to be readily usable. If a library containing the execution code is already available, the SIBCreator can also be applied in a bottom-up mode: methods from the libraries can be selected and the SIBCreator generates SIBs that are tailored to their invocation.

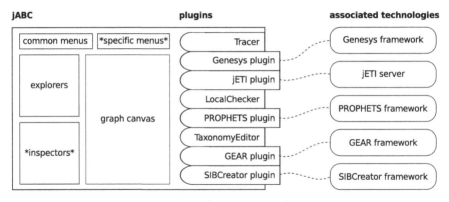

Fig. 2.4 Schematic overview of the jABC framework as used for Bio-jETI

Summarizing the above, Figure 2.4 gives a schematic overview of the basic parts of the jABC, the plugins and the associated technologies that have been used in the scope of the Bio-jETI incarnation of the framework. The jABC readily provides the menus that are common also to other programs (such as file loading, saving, project management, etc.) and the canvas and functionality for working with directed graphs, as well as menus and inspectors for different purposes. The "*" characters around the latter are to denote that these are the principal components to be extended and customized by the plugins in order to add specific functionality. Some of the plugins (Tracer, LocalChecker and TaxonomyEditor) work completely within the jABC framework, while the others integrate external functionality provided by associated technologies. For more detailed information about the architecture and plugin concept of the jABC framework, see [306, 229, 235].

Equipped with the plugins listed above, the jABC provides an intuitive environment for workflow development that hardly requires any classical programming skills. Workflow models are built graphically by constructing service compositions (SLGs) that orchestrate basic services (in the form of SIBs) along the flow of control. The atomic actions from which workflows are composed are provided by the workflow building blocks, hence the potential for the workflows is defined by the collection of available SIBs.

Technically, SIBs are Java classes that are (at least) marked with a `@SIBClass` annotation in order to be recognized by the jABC framework. Furthermore, a SIB class can contain definitions of parameters and labels for the outgoing branches, as well as documentation text and specific icons for the graphical representation. As indicated above, SIBs can furthermore implement the interfaces defined by jABC plugins to make use of their functionality. Most importantly, the Tracer and Genesys interfaces, which define methods for holding the execution code, are implemented by all service-providing SIBs. For an elaborate introduction to the SIB concept and technical implementation details, the reader is referred to [306, 229].

The basic jABC distribution provides rich SIB collections for commonly occurring tasks: SIBs that realize control-flow constructs like conditional branching, loops, and parallelism, but also libraries for working with lists and matrices and libraries for incorporating GUI elements. More precisely, these so-called *Common-SIBs* comprise the following functionality that is used within Bio-jETI:

- *Basic SIBs* for working with the execution context. For instance, this library provides SIBs for putting objects into the execution context, for accessing objects from the contexts, for evaluating conditions and for basic operations on character sequences.
- *Collection SIBs* for dealing with arrays, collections and maps. In addition to adding and removing objects from different kinds of collections, this library offers, for instance, SIBs that iterate over a collection object or perform specific operations like sorting a list.
- *GUI SIBs* for showing dialogs for user interaction. This library contains common dialogs, such as for message display, file selection, and the input of login data.
- *IO SIBs* for performing file-related tasks. The reading and writing of files are probably the most important and most frequently used functions that this library contains. However, also the other functionality it provides, such as browsing directories, executing console commands and zipping and un-zipping files are often useful.

Another set of standard SIBs in the scope of the Bio-jETI incarnation of the jABC framework are the SIBs that are commonly used when working with jETI services, which the jETI plugin provides:

- *jETI helper SIBs* provide basic functionality for loading data from different sources so that it can be transferred to the jETI server(s), as well as for retrieving and further processing the returned results.

The workflow developer can deliberately make use of the framework, the plugins and the readily provided SIBs as described above. In addition, concrete jABC workflow projects comprise:

1. *application-specific SIB libraries*, providing the functionality that is required for the developed workflows, but not provided by the standard set of SIBs,
2. PROPHETS *domain models* as described in detail in Section 2.3.2, comprising service and type taxonomies, semantic service annotations and constraints, and
3. the actual *SLGs* that implement the workflows.

Focusing on the application of the framework in the bioinformatics application domain, the development of workflow applications is in fact what constituted the main part of the work described in this book. This becomes evident in particular in Part II, where the selected bioinformatics application scenarios are described in detail, comprising the provisioning of specific SIB libraries, the definition of semantic domain models, and the actual workflow design.

2.1.2 The jETI Platform

The jETI platform [301, 202, 203] is an easy-to-use tool integration technology that is tailored to making file-based applications remotely accessible. It has been name-giver for Bio-jETI, since the integration of all kinds of existing analysis tools in order to provide them in a form that is easily accessible by workflow frameworks (like the jABC) is central to the rigorously service-oriented approach to bioinformatics workflow management that has been followed by this work.

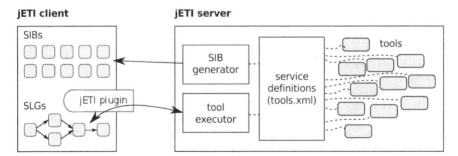

Fig. 2.5 Schematic overview of the jETI framework

Integration of services by means of jETI is convenient especially for tools that work on files as input and output data and have proper command line interfaces. This is, for instance, the case for many legacy applications, but also for many script-controlled tools. As illustrated by Figure 2.5, in jETI, the service provider maintains a jETI server that accesses (a collection of) tools on the one side, and on the other side provides an interface to the network. At runtime, the server receives service requests from a client (for instance, a jABC SLG) and forwards them to the actual applications, then collects the results and builds adequate response messages for the client. Conceptually similar to the Web Services' WSDL descriptions [69], relevant request parameters and the actual commands that are used by the server to execute the tools are maintained in an XML file (named `tools.xml`). This information is also used by the jETI server to generate the corresponding SIBs automatically.

Fig. 2.6 jETI Tool Server Configuration Interface

Since the jETI server provides an HTML-based tool configuration interface, it is not necessary to write the above-mentioned XML tool description file manually. This makes tool integration easy: the jETI server deals with most technical details, and the user only has to provide the information that is essential for the tool invocation. This comprises at least the path to the executable and required input and output parameters. Additionally, documentation texts and custom icons for the generated SIBs can be provided. Figure 2.6 gives an impression of the tool configuration interface: The start page (left) displays a list of all currently configured tools and provides links to, for instance, tool editors, server logs, and the SIB download site. The right side of the figure shows the definition of a mandatory input file parameter.

2.2   Constraint-Guarded Workflow Design

This section focuses on how Bio-jETI facilitates the analysis and verification of workflows via model checking. Model checking (cf., e.g., [70, Chapter 4] or [228]) provides a powerful mechanism to analyze and verify static aspects of (arbitrary) models. Generally speaking, it can be used to check whether a model M satisfies a property ϕ, usually written as

$$M \models \phi$$

where ϕ is expressed in terms of a modal or temporal logic. These logics provide powerful means to express workflow properties. In fact, a single formula can characterize an analyzed model up to bisimulation (cf., e.g., [292, 299]).

In model-based workflow development, model checking can help to detect problems already in the design phase. It is in particular useful to analyze properties of the whole workflow model, where syntax or type checking at the component level is not sufficient. Constraints are checked at modeling time, without execution of the process, which offers another range of addressable issues in addition to local validation and usual debugging methods. The list of properties against which the model is evaluated is easily extensible, since including a new constraint in the verification only requires to provide a modal or temporal formula expressing the property of interest.

In the following, Section 2.2.1 introduces the GEAR model checking plugin [35, 36], which allows for the model-wide evaluation of static properties (expressed in terms of modal or temporal specifications) directly within the jABC. Section 2.2.2 then deals with its application for the monitoring of user-level workflow constraints, before Section 2.2.3 addresses the use of model checking for the analysis of more technical properties.

2.2.1   Model Checking with GEAR

The SLGs of the jABC are mathematically analyzable objects that are directly amenable to formal analysis techniques, like model checking. More concretely, the directed graphs whose nodes represent basic services and whose edges define the flow of control can be interpreted as Kripke Transition Systems (cf. [228]). That is, they combine Kripke Structures (labeled states, unlabeled transitions, cf., e.g., [70, Chapter 2]) and Labeled Transition Systems (unlabeled states, labeled transitions, cf., e.g., [149]) into model structures where both states and transitions are labeled.

Basis for the model checking of workflows are the *atomic propositions*, simple properties that hold for single nodes of the workflow model. GEAR facilitates the annotation of SIB instances with arbitrary atomic propositions via a simple interface. Moreover, the names of model entities (such as SIBs, branches, and parameters) can directly be used as atomic propositions within the constraints.

GEAR basically uses the *Computation Tree Logic* (CTL) [70, Chapter 3] to formulate model checking constraints. CTL is a temporal, branching time logic designed to reason about models represented as directed graphs, and whose syntax can be described by the following BNF:

$$\phi ::= p \mid \neg\phi \mid \phi \vee \phi \mid \phi \wedge \phi \mid$$
$$AX(\phi) \mid EX(\phi) \mid AF(\phi) \mid EF(\phi) \mid AG(\phi) \mid EG(\phi) \mid$$
$$ASU(\phi, \phi) \mid ESU(\phi, \phi) \mid AWU(\phi, \phi) \mid EWU(\phi, \phi)$$

Thus, in addition to the operations and operands known from propositional logic, it comprises the modalities AX, EX, AF, EF, AG, EG, ASU, ESU, AWU and EWU. The As and Es are *path-quantifiers*, providing a universal (A) or existential (E) quantification over the paths beginning at a state. X, F, G, and SU and WU express *linear-time modalities* for the path. X expresses that ϕ must be valid in the *next* state, F specifies that ϕ must hold *finally*, and G requires ϕ to hold *generally* (always). SU (*strong until*) expresses that ϕ_1 has to be valid until ϕ_2, with requiring to ϕ_2 to hold finally. The *weak until* modality (WU) also requires ϕ_1 to be valid until ϕ_2, but does not require ϕ_2 to hold if ϕ_1 holds infinitely. A formal definition of the semantics can be found, for instance, in [70, Chapter 3] or [228].

Model checking in terms of CTL works on the Kripke Structure portion of the SLGs. In order to enable the formulation of model checking constraints with respect to particular branches, GEAR can also interpret the *box* and *diamond* operators known from Hennessy-Milner logic (HML) [130]:

$$[Trans]\phi \mid \langle Trans \rangle \phi$$

where *Trans* is a (possibly empty) set of labeled transitions (branches). The box operator expresses that ϕ must hold for all successors of a particular state that are reachable via one of the branches in *Trans* (universal quantification). Analogously, the diamond operator requires ϕ to hold for at least one *Trans*-successor (existential quantification).

GEAR extends these variants of CTL and HML further and includes additional overlined modalities that represent a backward view, that is, consider the paths that end at a given state. As an example, the formula:

$$\overline{AG}(\phi)$$

is fulfilled for all states where ϕ holds globally on all incoming paths.

Moreover, GEAR offers a variety of macros that provide composite operators or access to jABC-specific functionality.

2.2.2 Monitoring of User-Level Workflow Constraints

As detailed above, the GEAR model checking plugin allows for the global and continuous evaluation of (modal or temporal logic) constraints on the workflow model under development. This facilitates a constraint-guarded workflow development style, where workflow design is accompanied by the continuous evaluation of constraints that express domain-specific knowledge in terms of static properties of the workflow model. In particular, the constraints can capture semantic properties of the workflow models, such as their purposes, or rules of best practice. Similar to a spell checker in a text processing program, which hints at the violation of orthographic rules, perhaps already indicating a proposal for correction, the model checker notifies the workflow designer when the defined constraints are violated.

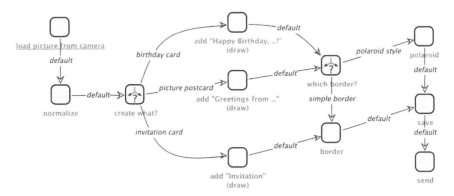

Fig. 2.7 Extended card creation workflow

For illustration of model checking with GEAR, consider the workflow model depicted in Figure 2.7, which extends the birthday card creation workflow example from Section 1.1.2: Instead of directly creating a birthday card from a picture by applying the **draw** and **polaroid** ImageMagick services, here the user can choose whether he wants to create a birthday card, a picture postcard or an invitation card. For the former two cases he can furthermore select whether to add a simple border or one in **polaroid**-style, before the resulting image is finally saved and sent.

In terms of GEAR's logic, it can now be verified, for instance, that:

- it is possible to create a card without using the **polaroid** effect:

$$EG(\neg \text{"polaroid"})$$

- after loading a photo from the camera, finally the (processed) image is saved:

$$\text{"load picture from camera"} \Rightarrow AF(\text{"save"})$$

- if the user has selected to create a birthday card, the suitable text is added:

$$[\text{"birthday card"}](AF(\text{"add happy birthday"}))$$

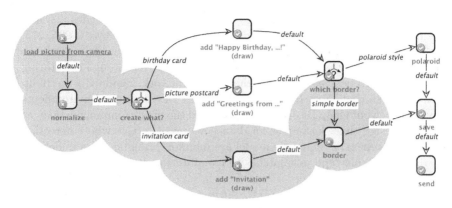

Fig. 2.8 Detection of a constraint violation via model checking

To visualize the model checking results directly within the workflow model, the GEAR plugin adds small overlay icons to all SIBs. As shown in Figure 2.8, green check mark icons are added to all SIBs for which the currently applied constraint is fulfilled, while the SIBs at which the constraint is not fulfilled are highlighted by red icons with a white cross. The figure shows the result of checking a constraint that requires that the **border** service must not be called unless the user has chosen this option, formally expressed as:

$$AWU(\neg\,\text{"border"}, \text{"which border?"} \wedge \langle\,\text{"simple border"}\rangle\text{true})$$

As the figure shows, this constraint is violated by the SIBs on the path that goes to the **border** SIB via the invitation-card branch of the workflow, which does not include the **which border?** SIB that lets the user make a choice.

Also the example constraint from Section 1.1.2, that in order to make a birthday card, the **draw** and **polaroid** services have to be applied to the picture can be expressed in terms of GEAR's logic:

$$[\text{"birthday card"}](AF(\text{"draw"}) \wedge AF(\text{"polaroid"}))$$

It does, however, not hold in the extended workflow model, either, as it is possible to create a birthday card with a simple border, which is in conflict with the constraint that explicitly demands the **polaroid**-style border.

Revisiting the spell checker analogy from the beginning of this section, note that the model checker's alerts should similarly be regarded as proposals for correction: finally, it is up to the workflow designer to decide whether the workflow model needs to be corrected or if instead the applied constraints have to be refined in order to "learn" to accept a new situation.

2.2.3   *Data-Flow Analysis via Model Checking*

On a somewhat more technical level, namely with respect to the flow of data within the workflow, the mature program analysis methodology that has been established in programming language compilers in the last decades [23] and that has been shown to be realizable via model checking [293, 294, 295, 273] can be applied. By providing a predefined set of desirable process properties to the model checker, it is for instance possible to achieve a thorough monitoring of safety and liveness properties within the framework. Similar to the built-in code checks that most Integrated (Software) Development Environments provide, this would help Bio-jETI users to avoid the most common mistakes at workflow design time. Examples for data-flow-related errors whose detection requires awareness of the whole model are manifold, ranging from undefined variables or simple type mismatches to computational gaps and incomplete processes.

As has been shown in [168, 171], *intra*procedural data-flow analysis via model checking can in fact be carried out directly in the jABC framework using the GEAR model checking plugin once atomic propositions are available that describe which data a SIB reads (uses) and writes (defines). For instance, an atomic proposition use(x) can be added to a SIB to express that a data item x is used by the service, or def(x) to state that it is written (defined). Based on such atomic propositions, it can now, for example, be checked that a service can not use an input x unless it has been defined. In terms of the GEAR syntax, this can be expressed as:

$$AWU(\neg\text{"use(x)"}, \text{"def(x)"})$$

This is already sufficient to ensure that the variable x has been defined, but does not say anything about type correctness. As described in greater detail in [174], it is therefore reasonable to extend the atomic propositions and the constraints in order to include the types of the involved variables. Accordingly, a formula like:

$$(\text{"use(x)"} \wedge \text{"type(x)=y"}) \Rightarrow \overline{ASU}(\neg\text{"def(x)"}, \text{"def(x)"} \wedge \text{"type(x)=y"})$$

can be used to express that if a service uses x of type y, x must have been defined before with precisely this type, without having been overwritten since.

It is of course possible to analyze other data-flow properties simply by providing the appropriate constraints. For a more comprehensive description of data-flow analysis via model checking in the jABC in general and on Bio-jETI workflows in particular, the reader is referred to [171] and [174], respectively. As shown in detail in [173, 174], model checking of data-flow properties can in fact help to detect common technical problems, like:

- Undefined data identifiers: a service uses a data item that has not been produced before.
- Missing computations: a workflow step is missing, so that a required resource is not fetched/produced.
- Type mismatches: a certain service is not able to work on the data format provided by its predecessor(s).

Solving these problems might require the introduction of further computational steps, for instance a series of conversion services in case of a data type mismatch. The automatic creation and insertion of such missing service sequences is in fact possible with the automatic workflow composition methods introduced in the next section.

Further on, *inter*procedural analyses could similarly be integrated via model checking of context-free/pushdown systems as described in [58, 59, 60]. In fact, it would even become possible to efficiently check fork-join parallel programs with procedures for the popular bitvector problems [153, 278]. This would allow one to elegantly capture the typical kind of parallelism that appears in workflows. It is envisaged to integrate these approaches into the jABC to form a workflow analysis framework in the spirit of the Fixpoint Analysis Machine [298], which combined a variety of intra- and interprocedural methods. Finally, the typical handling of data in workflows often leads to situations where one and the same object may have many syntactically different representations. It is therefore planned to investigate the impact of alias-sensitive analysis methods like the ones based on the Value Flow Graphs [300].

2.3   Constraint-Driven Workflow Design

The previous section discussed how the continuous evaluation of constraints via model checking facilitates a constraint-*guarded* workflow design style. Constraint-*driven* workflow design extends the application of constraints further, systematically using them for high-level descriptions of individual components and entire workflows, which can then be translated automatically into concrete workflows that conform to the constraints by design. This section introduces the functionality for (semi-) automatic, constraint-driven workflow design by which the jABC has been extended in the scope of this work. More precisely, it describes the PROPHETS (Process Realization and Optimization Platform using a Human-readable Expression of Temporal-logic Synthesis) plugin [234], which is the current implementation of the loose programming paradigm [178]. It was started by Stefan Naujokat in his Diploma thesis [232] and has since then been developed further in the scope of different application projects [233, 180, 179].

In accordance with the loose programming paradigm, PROPHETS simplifies workflow development in order to reach application experts without programming background. Therefore, the workflow designer neither needs to model fully executable workflows (as usually necessary) nor to formally specify a synthesis or planning problem in terms of some first-order or temporal logic. While behind the scenes the synthesis algorithm naturally requires formal specifications of the synthesis problem, PROPHETS hides this formal complexity from the user and replaces it by intuitive (graphical) modeling concepts. Thus, it integrates automatic service composition methodology into the jABC framework, but with little of its technical details being required for the user to know. In particular, PROPHETS features:

- graphically supported semantic domain modeling,
- loose specification within the workflow model,
- non-formal specification of constraints using natural-language templates,
- automatic concretization of the loose model into a concrete workflow, and
- automatic generation of model checking formulae.

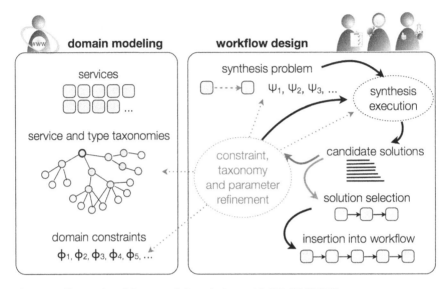

Fig. 2.9 Constraint-driven workflow design with PROPHETS

As Figure 2.9 shows, working with PROPHETS incorporates two major phases: *domain modeling* and *workflow design*. Whereas in the domain modeling phase PROPHETS is prepared for the application domain, the workflow design phase is dominated by the iterative "playing" with loose specification, synthesis, and constraint and parameter refinement. In particular, PROPHETS supports a very flexible way of expressing domain knowledge: it can either be specified during domain modeling (especially suitable for domain-specific constraints) or during the actual synthesis (especially suitable for problem-specific constraints). In this way, users can flexibly interact with the workflow development framework, collecting possible solutions and continuously refining the constraints according to any arising requirements.

In the following, Section 2.3.1 sketches the principles of the underlying synthesis method, before Sections 2.3.2 and 2.3.3 give a detailed introduction to PROPHETS by elaborating on the domain modeling and the workflow design phases, respectively.

2.3.1   The Synthesis Method

Generally, the term *process synthesis* is used to refer to techniques that construct workflows from sets of services according to logical specifications [196, 119]. The algorithm that is used by PROPHETS is based on the modal logic SLTL (Semantic Linear Time Logic) that combines relative time with descriptions and taxonomic classifications of types and services [305, 97, 98]. It takes two aspects into account: On the one hand, the workflow must ensure a valid execution regarding type correctness, on the other hand, the constraints specified by the workflow designer must be met.

The algorithm had already been implemented for the predecessors of the current jABC and jETI frameworks, that is, for the ABC and ETI platforms [98, 207], and was recently also used in the scope of the Semantic Web Services Challenge [248] for synthesizing mediators between different message formats [198, 159]. Only with the introduction of PROPHETS, however, the synthesis method has become conveniently accessible also for "normal" users: Background knowledge about the underlying method is not required for using the plugin, since all formal specifications that the synthesis algorithm needs are derived automatically from the intuitive, graphical specification mechanisms that PROPHETS provides.

Hence, knowledge of the internal details of the synthesis algorithm is not required in the scope of this book, and accordingly this section is restricted to sketching the basic ideas of how the algorithm works. Concretely, the following describes how the *synthesis universe* (which constitutes the search space in which the synthesis algorithm looks for solutions to the synthesis problem) is built from the provided domain model, how SLTL is used for the abstract workflow *specification formula*, and what the *synthesis algorithm* can finally derive from this information. For further details the reader is referred to [305].

Synthesis Universe

The synthesis method relies on behavioral service interface descriptions, that is, services are regarded as transformations that perform particular actions on the available data. Concretely, each service interface description must characterize the service by means of four subsets of the set of all data types:

- *USE* are the types that must be available before execution of the service (i.e. the input types of the service),
- *FORBID* describes a set of types that must not be available before execution of the service,
- *GEN* is the set of types that are created by the execution of the services (i.e. the output types of the service),
- *KILL* defines those types that are destroyed and therefore removed from the set of types that were available prior to execution of the service.

The synthesis algorithm then combines service descriptions in terms of these sets into the synthesis universe, that is, an abstract representation of all possible solutions that contains all service sequences that are valid (type-correct) executions, without taking into account any problem-specific information. The synthesis universe is in essence an automaton that connects states with edges according to available services. While each state represents a subset of all types (abstract and concrete), the connecting edges perform the transition on those types, according to input and output specifications of services. Every path in this automaton, starting from a state that represents the initially available data, constitutes an executable service sequence. As the synthesis universe is usually very large, it is not explicitly generated from the domain definition, but specified by sets of constraints that are evaluated on the fly during the synthesis process.

Note that the synthesis method uses *taxonomies* to define semantic classifications of types and services. Taxonomies are simple ontologies that relate entities in terms of *is-a* relations and thus allow for the hierarchical structuring of the involved types and services. The actually available services and types are named *concrete*, whereas semantic classifications are named *abstract*. The taxonomies are considered by the synthesis algorithm when constructing the synthesis universe and when evaluating type and service constraints.

Formally, as for instance defined in [178], the synthesis universe is a triple $(T, S_c, Trans)$, where

- T is a set of concrete and abstract types
- S_c is a set of concrete services
- $Trans = \{(t, s, t')\}$ is a set of transitions where $t, t' \subseteq T$ and $s \in S_c$.

Note that the synthesis universe only contains the set of concrete services (S_c), as it contains real executable paths. Abstract services (S_a) are solely used as part of the specification formula (see below). The set of all services is denoted as $S = S_a \cup S_c$. The sets of types T_c, T_a and T are defined likewise.

With the definitions of USE, GEN, $FORBID$, and $KILL$, a service $s \in S_c$ can be defined as a transformation on the power set of types as follows:

$$s : 2^T \to 2^T$$
$$t \mapsto (t \setminus KILL(s)) \cup GEN(s)$$

For s to be admissible in a type state $t \subseteq T$, the following conditions must be met: $USE(s) \subseteq t$ and $FORBID(s) \cap t = \emptyset$. At last, the synthesis universe can be constructed from the service definitions as follows: for each $t \subseteq T$, a state in the universe is created. The transition (t, s, t') is added to $Trans$ iff s is admissible in t and $t' = (t \setminus KILL(s)) \cup GEN(s)$.

Specification Formula

The specification formula describes all sequences of services that meet the individual workflow specification, but without taking care of actual executability concerns. It is given declaratively as a formula in SLTL, a semantically enriched version of the commonly known propositional linear-time logic (PLTL) that is focused on finite paths. The syntax of SLTL is defined by the following BNF, where t_c and s_c express type and service constraints, respectively:

$$\phi ::= true \mid t_c \mid \neg\phi \mid \phi \wedge \phi \mid \langle s_c \rangle \phi \mid G\phi \mid \phi U\phi$$

Thus, SLTL combines static, dynamic, and temporal constraints. The static constraints are the taxonomic expressions (boolean connectives) over the types or classes of the type taxonomy. Analogously, the dynamic constraints are the taxonomic expressions over the services or classes of the service taxonomy. The temporal constraints are covered by the modal structure of the logic, suitable to express the ordering constraints:

- $\langle s_c \rangle \phi$ states that ϕ must hold in the successor state, and that it must be reachable with service constraint s_c.
- G expresses that ϕ must hold generally.
- U specifies that ϕ_1 has to be valid until ϕ_2 finally holds.

In addition to the operators defined above, it is convenient to derive further ones from these basic constructs, such as the common boolean operators (disjunction, implication, etc.), the eventually operator $F\phi =_{def} true \; U \; \phi$, or the weak until operator $\phi \; WU \; \psi =_{def} (\phi \; U \; \psi) \vee G(\phi)$. Furthermore, the next operator $X \; \phi$ is often used as an abbreviation of $\langle true \rangle \phi$. A complete formal definition of the semantics of SLTL can be found, for instance, in [301, 178].

The distinctive feature of SLTL formulae is that they cover two dimensions:

1. The *horizontal* dimension, covered by the modalities that describe aspects of relative time, addresses the workflow model, including its loosely specified parts, and deals with the actual service sequences.
2. The *vertical* dimension evaluates taxonomic expressions over types and services, allowing for the usage of abstract type and service descriptions within the specifications.

Both kinds of constraints can deliberately be combined in order to express more complex intents about the workflows. This allows for a very flexible fine-tuning of the workflow specifications.

The specification formula that is finally used as input for the synthesis algorithm is simply a conjunction of all available SLTL constraints, comprising:

- The definition of the *start condition(s)* of the workflow (i.e., the set of data types that is available at the beginning).
- The definition of the *end condition(s)* of the workflow (i.e., the set of data types that must be available at the end of the synthesized workflow).
- The set of available *workflow constraints*.

Synthesis Algorithm

The synthesis algorithm then interprets the SLTL formula that specifies the synthesis problem over paths of the synthesis universe, that is, it searches the synthesis universe for paths that satisfy the given formula. The algorithm is based on a parallel evaluation of the synthesis universe and the formula (for details on the algorithm, cf. [305]). It computes all service compositions that satisfy the given specification.

2.3.2   Domain Modeling

During the *domain modeling* phase, as shown Figure 2.9, PROPHETS is prepared for the application domain. Roughly speaking, the domain modeling involves everything that is required prior to domain-specific workflow development, such as service integration and providing meta-information about the services and data types of the application domain. This activity requires an overall understanding of the specific domain, as well as explicit knowledge of the available resources, and in particular concerning their interplay. Thus, the task of a domain modeler, whose role it is to formalize this knowledge in terms of ontologies and other kinds of formalisms, is different from the task of the workflow developer, who exploits the domain model when working with the readily customized workflow framework.

More precisely, setting up a domain for PROPHETS involves four major steps (as depicted on the left hand side of Figure 2.9):

1. Integration of services.
2. Definition of the available services and data types, and hierarchical structuring of the domain by semantic classification of types and services via taxonomies.
3. Interface description of services in terms of their input/output behavior.
4. Formulation of domain-specific constraints, such as ordering constraints, general compatibility information or dependencies between services.

In the jABC context, service integration simply means the provisioning of SIB libraries as explained earlier (cf. Section 2.1). The other three aspects of domain modeling are detailed in the following.

Service and Type Taxonomies

PROPHETS supports the definition of type and service taxonomies as used by the synthesis algorithm. Technically, the taxonomies are stored in OWL format [276], where the concept *Thing* denotes the most general type or service and the OWL classes represent abstract classifications. The concrete types and services of the domain are then represented as individuals that are related to one or more of those classifications by *instance-of* relations.

PROPHETS uses OntEd [216], a jABC plugin for the visualization and editing of OWL ontologies, as a simple, built-in editor for the type and service taxonomies.

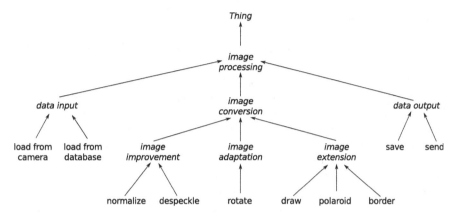

Fig. 2.10 Exemplary service taxonomy

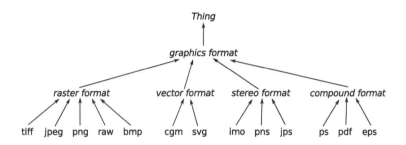

Fig. 2.11 Exemplary type taxonomy

Similar to the exemplary service and type classification schemes of Section 1.1.2, Figures 2.10 and 2.11 depict possible type and service taxonomies, respectively, as they may be defined for the birthday card creation workflow example. The generic root *Thing* and the abstract classes in the taxonomies are written in italics, while the concrete service and type instances are set in normal font. The service taxonomy distinguishes the *images processing* services in the domain further into *data input* (comprising the services for loading data from a camera and from a database), *data output* (saving and sending files) and *image conversion services*. The latter defines again three subclasses, namely *image improvement* (**normalize, despeckle**), *image adaptation* (**rotate**) and *image extension* (**draw, polaroid, border**). The type taxonomy defines four sub-categories for *graphics formats*, namely *raster format* (covering the **tiff, jpeg, png, raw** and **bmp** file types), *vector format* (**cgm, svg**), *stereo format* (**imo, pns, jps**) and *compound format* (**ps, pdf, eps**).

Note that these taxonomies are also taken as the basis in the following when further examples for illustrating the loose programming approach are given.

Service Interface Descriptions

As detailed in Section 2.3.1, each service interface description in PROPHETS must characterize the service by means of its USE, GEN, KILL and FOR-BID sets. In order to enable thorough abstraction from the concrete service implementations, services and data types are represented by *symbolic names* throughout the framework. Thus, service descriptions in terms of the domain model are clearly decoupled from the technical service specifications and implementations, so that any kind of heterogeneous resource at any location can be integrated, and there is no restriction to readily semantically annotated services of a particular platform. Moreover, as all service meta-information is stored as a separate file within the current project's directory, it is possible to use a specialized nomenclature for different jABC projects, even when the underlying services are the same.

Table 2.1 Exemplary service interface descriptions

Service	USE (input types)	GEN (output types)
border	*raster format*	png
draw	*raster format*	png
load from camera		raw
load from database		svg
normalize	*raster format*	png
rotate	*raster format*	png
save	*graphics format*	
send	*raster format*	

To give an example, Table 2.1 lists a couple of services from the birthday card creation workflow scenario along with their interface annotation in terms of USE and GEN sets, that is, their input and output data types. Note that italics are used to denote when this information is given in terms of an abstract classification from the type taxonomy, which typically means that the respective entry represents different possible concrete data types.

The data loading services (for loading a **raw** file from a camera and for loading a **svg** file from a database) require no inputs and thus have an empty USE set, whereas their GEN sets simply consists of the output data they produce. The ImageMagick-based picture processing services accept all kinds of *raster format* images as input and generate **png** files as output. Finally, the **send** and **save** can process *raster format* and arbitrary *graphics format* inputs, respectively, but generate no output data and have empty GEN sets.

Apparently, none of these services defines any entries for the FORBID and KILL sets. This is in fact representative for all application domains considered in this book, which primarily work on self-contained data processing services that simply consume and produce data, so that no further behavioral aspects of the individual services have to be taken into account.

In addition to the input/output characterization in terms of symbolic type names, a PROPHETS service interface description comprises technical information like the reference to the SIB that implements the service, the SIB branch to which the interface description applies, and a mapping from symbolic types to the corresponding concrete SIB parameters (to enable automatic parameter configuration). Furthermore, it is possible to provide default values for SIB parameters that are not included in the semantic interface descriptions of the domain model, and which are consequently not be configured by the synthesis plugin automatically. However, with appropriate default values provided, executable workflows can be obtained also in these cases.

Note that the current implementation of the synthesis framework only considers sets of data *types* for input/output characterization of the services in the domain, and does not distinguish between concrete *instances* in terms of the services' actual parameters. That is, even if several parameters of the same type are involved, the corresponding sets contains the type exactly once, and thus the synthesis framework is not able to tell the individual instances apart. This can cause ambiguities, especially with respect to inserting and instantiating services that have more than one parameter of a particular type.

For instance, consider an image processing service that combines two (or more) pictures of the same format, say jpg, into one. The USE set of this service only comprises the type jpg. Consequently, the synthesis regards this service as executable as soon as one jpg data item is available, although actually one (or more) additional instances of this type are required. Similarly, when transferring a synthesis solution (which is given as a sequence of states that are constituted by the sets of available data types and connected by services that consume and produce these types) into a concrete SLG, the instantiation of the involved SIBs can face ambiguities. As an example, the availability of a particular data type at a state does not tell whether it refers to one or more instances of this type, and in the latter case, to which. That is, if the type jpg is available, but more than one of the (input) parameters of the respective SIB are of this type, then it is not clear how the parameters have to be instantiated. As another example, the occurrence of a data type in several states of the synthesis solution alone does not reveal whether it refers to the same or to different instances of the type. That is, a jpg at two different states may refer to exactly the same image, or to different ones.

Currently, PROPHETS applies a simple heuristics for the parameter instantiation, which basically aims at establishing short-distance data flow connections, that is, data items are rather passed between subsequent SIBs than across longer service sequences. In case of remaining ambiguities, simply the first possibility that is encountered is used. Fortunately, this strategy has turned out to be sufficient for the most part of the considered application scenarios, so that only in very few, special cases the parameters of the SIBs had to be (re-) configured manually in order to obtain fully executable workflows. Nevertheless, developing more sophisticated heuristics or even enabling the framework to perform also instance-based workflow synthesis and thus properly overcome this issue is a central subject of future work (cf. Section 9.2.1).

Domain-Specific Constraints

Fig. 2.12 PROPHETS' Domain Constraints Editor

The domain-specific constraints must be formalized appropriately, that is, expressed in SLTL, which is the temporal logic underlying PROPHETS' synthesis method (cf. Section 2.3.1). As understanding and writing SLTL constraints can be extremely difficult for users without background in formal methods and inconvenient also for those who are familiar with temporal logics, PROPHETS supports constraint formulation by means natural-language templates. That is, it provides a number of predefined constraints along with a natural-language equivalent of the SLTL formula that is displayed by the constraint editor. Figure 2.12 shows PROPHETS' constraint editor in action: the available constraint templates are displayed in a list in the upper part of the editor window, all template instances are displayed in the lower part. As the figure shows, placeholders are used for the services and types within the templates. The replacements for the placeholders can simply be selected from drop-down lists which the editor fills with all valid options (i.e., all concrete and abstract services or types as defined by the interface descriptions and the respective taxonomies).

Several of the templates that are provided by PROPHETS by default are the result of experiences with the application of PROPHETS to the bioinformatics domain. As the constraint templates are defined in a simple XML file, advanced users can easily extend the list to the needs of the specific domain. Currently, the following templates are available (template name, description/cloze text, SLTL representation):

- **service avoidance**:
  Avoid the service s.
  $G(\neg\langle s\rangle true)$
- **conditional service avoidance**:
  If service s_1 is used, avoid the service s_2 subsequently.
  $G(\langle s_1\rangle true \Rightarrow X(G(\neg\langle s_2\rangle true)))$
- **mutual exclusion of services**
  At most one of the services s_1 and s_2 may be used.
  $\neg(F(\langle s_1\rangle true) \wedge F(\langle s_2\rangle true))$
- **service redundancy avoidance**:
  Do not use service s more than once.
  $G(\langle s\rangle true \Rightarrow X(G(\neg\langle s\rangle true)))$
- **service enforcement**:
  Enforce the use of service s.
  $F(\langle s\rangle true)$
- **conditional service enforcement**:
  If service s_1 is used, enforce the use of service s_2 subsequently.
  $G(\langle s_1\rangle true \Rightarrow X(F(\langle s_2\rangle true)))$
- **service succession**:
  If service s_1 is used, service s_2 has to be used next.
  $G(\langle s_1\rangle true \Rightarrow X(\langle s_2\rangle true))$
- **service dependency**:
  Service s_1 depends on service s_2 (i.e., service s_1 can only be used after s_2).
  $\neg\langle s_1\rangle true \text{ WU } \langle s_2\rangle true$
- **final service**
  Use service s as last service in the solution.
  $F(\langle s\rangle true) \wedge G(\neg(\langle s\rangle true) \vee \neg(\langle\rangle\langle\rangle true))$
- **type avoidance**:
  Avoid the type t.
  $G(\neg t)$
- **type enforcement**:
  Enforce the existence of type t.
  $F(t)$

- **mutual exclusion of types**:
  At most one of the types t_1 and t_2 may exist.
  $\neg(F(t_1) \wedge (F(t_2)))$

Considering again the birthday card creation workflow scenario, the following lists some exemplary essential characteristics of the application that should be covered by the domain constraints, and that can in fact easily be expressed using templates from above:

- Enforce the use of `draw`.
  Enforce the use of `polaroid`.
- At most one of `border` and `polaroid` may be used.
- Do not use *image improvement* more than once.

As indicated earlier, the domain model can not only be used as basis for workflow synthesis, but also be applied for enhancing the workflow development support via model checking as described in Section 2.2. While the model checking is not necessary for workflow parts that were produced by the synthesis (as it takes care of meeting the constraints automatically), it can be useful in case that the workflow developer changes the synthesized parts later. On the one hand, domain constraints can be re-used as global workflow properties and evaluated by the model checker. On the other hand, model checking constraints can be derived from the service interface descriptions that allow for automatic checking of type consistency within a model using the input and output specifications (cf. Section 2.2.3). To achieve this kind of global type monitoring, PROPHETS generates the following formula structure for every type $t \in T$ automatically:

$$\left(\bigwedge_{u \in USE(t)} \neg u \right) \quad \mathbf{WU} \quad \left(\bigvee_{g \in GEN(t)} g \right)$$

where $USE(t)$ denotes the services that have t as input (use) and $GEN(t)$ the ones that have t as output (gen). Intuitively, the formula expresses that no service that uses t must be called before a service has been called that generates t.

2.3.3   Workflow Design with Loose Specification

In the actual *workflow design* phase, the workflow designer benefits from the domain model that has been set up according to his needs, referring to services and data types using familiar terminology. As detailed in the following, the concept of *loose specification* is central to workflow design with PROPHETS: it is possible to mark branches between SIBs as loosely specified to denote that they are abstract representations of the corresponding part of the workflow, and that their translation into a readily executable workflow is left to the synthesis framework.

Loose Specification

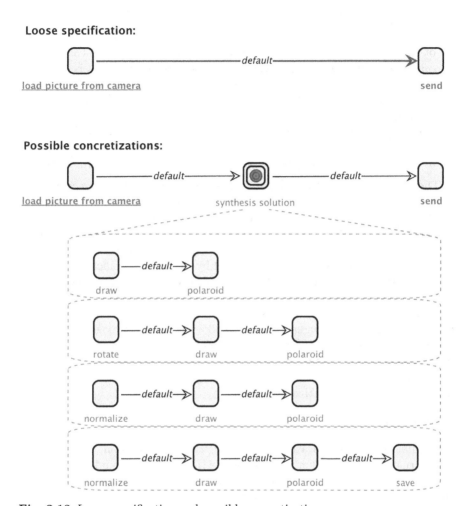

Fig. 2.13 Loose specification and possible concretizations

As Figure 2.13 (top) shows, loosely specified branches (here the default branch between the SIBs `load picture from camera` and `send`) are colored red by the PROPHETS plugin in order to be distinguishable from other branches. Intuitively, a loose branch represents *all* sequences of services that would constitute a valid connection between the respective SIBs. In order to form valid inputs for the synthesis algorithm, PROPHETS translates loose branches into SLTL specification formulae that comprise all available workflow constraints (cf. Section 2.3.1):

- The *start conditions*, that is, the set of data types that is available at the beginning, are determined according to the preceding services using data-flow analysis methods.
- The *end conditions*, that is, the set of data types that have to be available at the end of the synthesized workflow are derived from input types of the SIB at the destination of the loose branch. These so-called *goal constraints* can be *terminating* (i.e., the search for solutions is aborted once the required types are available) or *non-terminating* (i.e., the search for further solutions continues beyond the state where goal constraint has been fulfilled for the first time).
- The domain-specific *workflow constraints* as provided by the domain model and the problem-specific constraints as provided by the workflow designer (see below) are also simply added to the specification formula.

The result of the synthesis algorithm is then in fact set of all constraint-conform, executable concretizations of a loosely specified branch. Accordingly, the four exemplary concretizations that are shown in Figure 2.13 (bottom), where the `draw` and `polaroid`, and additionally also other services from the corresponding domain model are inserted, are only some of the many possible solutions for the synthesis problem defined by the loose specification above.

Note that although the examples applied in this book comprise only single loose branches to simplify matters, it is possible to mark arbitrarily many branches in a workflow model as loosely specified. Then, the synthesis is carried out separately for each branch, following their topological order in the workflow model. As detailed in Section 9.2.1, improvements with respect to precise constraint scoping and treatment of second-order effects (that occur when the workflow model evolves during the synthesis) are however necessary in order to fully support this feature.

Synthesis Execution

Depending on the configuration of the PROPHETS installation, the synthesis is performed without further user interaction (i.e., in "silent" mode, where simply the shortest solution is taken), or with different intermediate steps where the user can further customize synthesis inputs and parameters, and finally select an adequate result or refine the constraints in order to narrow the set of solutions. In the scope of this work, a PROPHETS configuration has been used that takes the user through the following synthesis steps, as summarized in Figure 2.14:

1. *Edit constraints*
   Displays the domain-specific constraints that are already provided by the domain model, which can be changed or removed. Additional *problem-specific* constraints can be provided that narrow the workflow designer's intentions further (using the template editor described earlier). As Figure 2.9 indicates, these problem-specific constraints are essential when

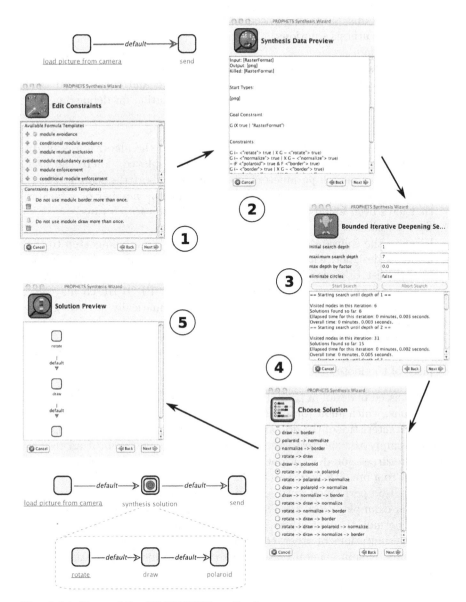

Fig. 2.14 Steps of the PROPHETS synthesis process

constraints and parameters need to be refined in order to improve the obtained synthesis solutions.

2. *Synthesis data preview*

Displays the domain information (service interfaces and taxonomies) and the specification formula (workflow specification and constraints) that will be applied. This step provides useful information for debugging purposes.

3. *Search*

After the synthesis data has been processed and the algorithm is ready to search for solutions, the search dialog is displayed. There the user can configure search parameters, and start and abort the search. Status information, like the current search depth and the number of solutions that have been found so far, is displayed while the search runs.

4. *Solution chooser*

Displays the solutions that have been found in the previous step to the workflow designer, who can then directly choose an appropriate solution from the list, or decide to return to the first step to refine the specification and restart the synthesis process. Alternatively, PROPHETS may also be configured to select one of the solutions automatically according to a particular cost function.

5. *Solution preview*

Shows the sequence of SIBs that will be inserted into the workflow in place of the loose branch.

As discussed in detail in [233], PROPHETS provides a variety of configuration options, which allow for a very specific tailoring of the synthesis process to the characteristics of the domain model. For instance, it can be configured whether simply the shortest synthesis solution shall be taken, if all results shall be returned, or if an iterative-deepening depth-first search (which is equivalent to a breadth-first search, but more space-efficient, cf. [255, Section 3.7.3]) shall be performed until a particular search depth has been reached. As another example, PROPHETS can be configured to ignore solutions that are mere permutations of others. Since often many solutions are found that are only permutations of other, equivalent service sequences that lead to the same results, it is in general reasonable to remove those permutations from the set of solutions that is presented to the user.

As of recently, PROPHETS also supports the definition of custom synthesis processes and thus provides a simple way of extending the plugin with further functionality. Since these processes are themselves modeled as workflows in the jABC, it is even possible to apply PROPHETS for their (semi-) automatic composition, and thus to synthesize synthesis processes.

Summarizing, as sketched in Figure 2.9, the workflow design phase constitutes the main working area of PROPHETS. Starting from a loosely specified workflow (top left corner of workflow design box in the figure) an iterative refinement of (workflow) constraints and synthesis parameters is applied until the user chooses one appropriate solution. If during the experimentation with the loose workflow refinement further general domain-specific patterns are detected, domain-specific constraints might be added to the domain model in order to, e.g., exclude domain-wide unsatisfactory solutions. Thus, PROPHETS supports both the more general domain-wide tailoring and the instance-oriented workflow-specific refinement of the synthesis environment.

Part II

Applications

3

Phylogenetic Analysis Workflows

This first example scenario is concerned with phylogenetic analyses. As they are comparatively easy to understand (also for non-biologists) and there is also a plethora of easy-to-use software tools available for the individual analysis steps, phylogenetic analyses have become a frequently used, quasi-standard application for illustrating bioinformatics workflow technology (cf., e.g., [256, 134, 155, 265, 165]). Furthermore, the annotation with semantic meta-data is particularly advanced for this discipline [179], which is advantageous for the application of the constraint-driven workflow design methodology.

3.1 Background: Phylogenetics

The term *phylogenetics* refers to the analysis of evolutionary relationships between different groups of organisms [279, p. 267]. While in former times phylogenetic analyses were based on the examination of morphological and physiological characters (the *phenotype*), modern molecular biology makes it possible to take also genetic information (the *genotype*) into account. The major carriers of genetic information are the *DNA* (deoxyribonucleic acid) molecules, which are typically present in the form of a *double helix*. The DNA double helix consists of two complementary nucleic acid strands that are held together by hydrogen bonds between the bases *adenine* (A), *cytosine* (C), *guanine* (G), and *thymine* (T). These *base pairs* (bp) hold together the two strands much like the rungs of a rope ladder. The *sequence* of the bases in a nucleic acid strand constitutes the information that is carried by the DNA molecule, that is, the individual *genes* that control specific features of heredity. During cell division the DNA is replicated so that the offspring obtains an identical copy of the genetic information. Another form of passing genetic information to the next generation is DNA recombination, that is, the assembly of genetic information from different sources into new sets of hereditary information, which is a major characteristic of sexual reproduction and results in greater genetic diversity.

According to the widely accepted theory of evolution (for which Charles Darwin laid the foundations in his 1859 book "On the Origin of Species" [80], and which was combined with the Mendelian knowledge of inheritance [218] in the 1930s), life evolves by means of natural selection, mutations and genetic drift [164]:

- *Natural selection* is a natural process that results in the survival and reproductive success of those individuals or groups that are best adjusted to their environment (the often quoted "survival of the fittest"). This process leads to the selection of genetic qualities that are best suited to a particular environment.
- *Mutations* are changes in the genome that occur spontaneously or due to the influence of mutagens such as UV light or chemicals. They lead to permanent loss (*deletion*), exchange (*substitution*) or addition (*insertion*) of bases in DNA sequences.
- *Genetic drift*, in contrast to natural selection, refers to random changes in the frequency of genes within populations. Over time, these changes lead to the preservation or extinction of particular genes.

In short, evolution leaves its marks in the genomes, and can be tracked by examination of molecular sequences.

3.1.1   Phylogenetic Trees

Phylogenetic trees are used to represent relationships between organisms graphically [279, p. 267]. Similar to family trees that show the relationships between

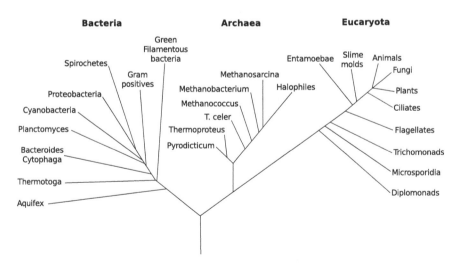

Fig. 3.1 Universal phylogenetic tree of life (following [346])

different members of a family, they depict the phylogenetic evolution of living organisms and the origin of species over the course of time. The exemplary "universal tree of life" depicted in Figure 3.1 shows a phylogenetic tree that visualizes the separation of the realms of bacteria, archaea, and eukaryotes [346]. As it was inferred from comparative analysis of ribosomal RNA (rRNA) sequences, it differs in details such as exact branching points and the exact position of the root (the suspected common ancestor) from trees that have been constructed based on other genes. The general picture, however, remains the same. Note that humankind is located within the "Animals" branch of the eukaryotes (in the upper right corner of the figure), and that the vast majority of species are microorganisms that are too small to be seen by the naked eye.

In the daily routine of molecular biology research, however, the phylogenetic relationships between different strains of particular organisms play a more important role than the universal tree of life. During the 2011 EHEC outbreak in Germany, for instance, phylogenetic methods were applied to analyze the E. coli strains that caused the severe hemolytic-uremic syndrome (HUS) [21]: Samples of the infectious variant of the normally harmless and vital intestinal bacteria were isolated and their genomes sequenced. A phylongetic analysis (incorporating a phylogenetic tree constructed from known E. coli strains and the newly sequenced ones) revealed that the new strains were in fact closely related to a known pathogenic strain. Subsequently, additional knowledge about the new strain could be derived from more detailed comparisons with the closely related strains. Collecting such comprehensive knowledge about the biological characteristics of a pathogenic organism is of course of academic interest, but furthermore essential for the development of effective treatments and medication.

3.1.2 Sequence Alignments

Sequence alignments (cf., e.g., [254, Chapter 6]) try to establish correspondences between the bases or codons of DNA, RNA, or amino acid sequences. They are used in phylogenetics to identify similarities between sequences which result from the existence of a common ancestor. Other applications of alignments are the matching of sequences against databases, the detection of overlaps and subsequences, the prediction of secondary and tertiary structures of proteins, as well as the prediction of active or functionally important sites of proteins.

```
TGGAC-CATTT (s₁)
TGCACGCATAT (s₂)
** ** *** *
```

Fig. 3.2 Simple pairwise nucleotide sequence alignment

Figure 3.2 shows a possible alignment for the nucleotide sequences TG-GACCATTT (s_1) and TGCACGCATAT (s_2). The asterisks indicate exact matches between bases, while the whitespace characters stand for mismatches (probably substitutions that occurred in the evolutionary process). The special gap symbol "-" represents potential insertions and deletions. Accordingly, the evolutionary history between the sequences s_1 and s_2 might be that two base substitutions (G to C at the third base, T to A two positions to the last base), and an insertion of a G into the central CC or a deletion of the G from the central CGC occurred.

```
CKAGTALWKVLQLNDSYDLDKHLDIKQYTHKLQQELQSFKVDLKDLDLLN
CKAGAVLWKVLQLNDSYDLDKHLDIKQYTHKIQQELQSFQVDLKELDLLS
CKEGAALWTVLQLNDSYDLEEHLDINQYTNKLRQELQSLKVDTQSLDLLS
CKEGAALWTVLQLNDSYDLEEHLDINQYTNKLRQELQSLKVDTQSLDLLS
----------------YLDQFVDLNRLYADLRAQAE-------------
                *::,.:*:::    .:: : :
```

Fig. 3.3 Simple multiple protein sequence alignment

A slightly more complex alignment is shown in Figure 3.3, which depicts a part of an alignment of multiple protein sequences. As in the Figure 3.2, the asterisk marks exact matches and the whitespace character marks mismatches. Additionally, the colon symbol is used to indicate conserved substitutions (i.e., replacements of amino acids by other ones that have similar chemical properties) while a single dot indicates semi-conserved substitutions (i.e., substitutions by an amino acid that has a similar spatial structure, but that does not have the same chemical properties).

The quality of an alignment can be expressed by a *score*, that is, a numerical value that is the sum of the scores of all positions in the alignment. For instance, simply assume the score for a match to be +1, for a mismatch 0 and for a gap -1, then the score for the example given in Figure 3.2 would be 7. Given a particular scoring function, the alignment that yields the highest score under these conditions is called the optimal alignment. Choosing appropriate scoring strategies is crucial in order to obtain reliable results. Roughly speaking, scoring functions for nucleotide sequences (consisting of four bases) are fundamentally different from those for amino acid sequences (20 amino acids that are encoded by base triples), and taking into account further aspects like evolutionary time or insertion/deletion probabilities makes the task no easier. Although the community has developed a set of standard scoring methods for common purposes (like, e.g., the PAM substitution matrix [81] or the BLOSUM matrices [129], it is often necessary to adjust the scoring scheme to the specific present case.

Pairwise sequence alignments are usually computed by dynamic programming methods, such as the Needleman-Wunsch alignment algorithm [236] or the Smith-Waterman algorithm [288]. In realistic experiments, however, it is often necessary to compare more than two sequences and hence to compute multiple sequence alignments. Applying the aforementioned methods for the alignment of multiple sequences becomes numerically intractable very fast, since it corresponds to searches through high-dimensional spaces instead of in just two dimensions. Consequently, algorithms are used that follow reliable heuristic strategies. The popular ClustalW [320], for instance, computes fast pairwise alignments of the input sequences in order to establish a so-called guide-tree, which is then used to settle the order in which the multiple alignment is successively assembled from the sequences (cf. Section 3.1.3).

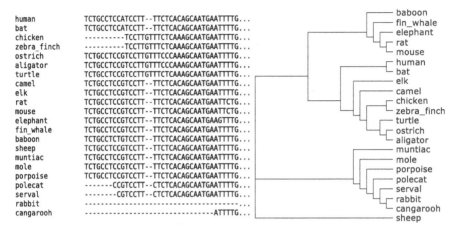

Fig. 3.4 Multiple sequence alignment and derived phylogenetic tree

Sequence alignments define *distances* between sequences. Roughly speaking, high sequence identity suggests that the sequences in question have a comparatively young most recent common ancestor (i.e., a short distance), while low identity suggests that the divergence is more ancient (a longer distance). Figure 3.4 gives an example of a (part of a) multiple sequence alignment and a derived phylogenetic tree. There are a number of distance-based methods for the construction of phylogenetic trees, among the most popular are the UPGMA algorithm [221] and the Neighbor-Joining method [268]. A detailed elaboration on this topic would go beyond the scope of this book, for understanding the presented examples is it sufficient to know that multiple sequence alignments provide one possible basis for the estimation of phylogenetic trees.

3.1.3   *ClustalW*

ClustalW [320] is the probably most popular multiple sequence alignment program. The algorithm behind it utilizes the fact that similar sequences are usually homologous [279, p. 81] and computes a multiple sequence alignment in three major steps:

1. Compute pairwise alignments for all sequence pairs.

2. Generate a phylogeny that represents the relationships between the sequences (the so-called *guide tree*) based on the scores of the pairwise alignments.
3. Carry out the multiple alignment successively according to the guide tree, starting with the sequences that are classified by the guide tree as most similar.

The algorithm can furthermore directly generate phylogenetic trees for the alignments, since it calculates the evolutionary distances between the sequences during the alignment computation.

ClustalW is actually only part of the Clustal program [182], which comprises ClustalW as a command line interface and ClustalX as a graphical user interface. The software accepts a wide range of sequence formats as input, and likewise supports a number of different output formats. It maintains default settings for the main input parameters, so that users can directly start aligning sequences. In particular, however, it is often useful to change parameters like the gap opening penalty and the gap extension penalty in order to obtain better alignment results.

Fig. 3.5 ClustalW execution via the command line interface

Figure 3.5 gives an impression of the ClustalW command line interface: After starting the program, the user is shown a list of functions to select from. Usually the first thing to do in a ClustalW session is to load the input sequences by selecting 1 and then entering the respective file name. Once the sequences have been provided, the user is shown the list of functions again. This time he can, for instance, choose 2 (Multiple Alignments) in order to compute a multiple sequence alignment from the input sequences. ClustalW prints elaborate status information during the algorithm execution, informing about the current state of execution (pairwise alignments, guide tree, multiple alignment) and about key numbers like the number of groups or the alignment score. Finally, also the resulting multiple alignment is displayed.

In addition to using the interactive command line interface, it is possible to run ClustalW in a non-interactive command line mode, that is, the input files and parameters are already specified in the command line call, and the result is printed on the console or written into a separate file. Altogether, ClustalW provides a quite "atomic" (i.e. small and self-contained) functional unit with well-defined inputs and outputs, which can be used easily in different contexts. Therefore it is also often provided as web service, allowing easy programmatic access also via the internet.

3.2 Variations of a Multiple Sequence Alignment Workflow

This first example of bioinformatics workflows with Bio-jETI is woven around multiple sequence alignments as described in Section 3.1.2. It is based on a number of data retrieval and sequence analysis services that are provided by the DNA Data Bank of Japan (DDBJ) [223, 165]. This example illustrates in particular the agility of workflow design with Bio-jETI: workflows can easily be modified, adapted, customized and tested in its graphical user interface, and (parts of) workflows can be prepared and flexibly (re-) combined at the user level according to current analysis objectives. The example also demonstrates that user interaction can easily be included in the workflows by using the SIB libraries that are shipped with the jABC framework.

The following gives an overview of the DDBJ services and describes the implementation of a minimal alignment workflow in Bio-jETI, before presenting a number of possible variations of the basic workflow. For a more elaborate description of this example, the reader is referred to [172].

3.2.1 DDBJ Services

The DDBJ [145] is a major database provider in Asia that forms the International Nucleotide Sequence Database (INSD) collaboration [72] together with the NCBI's GenBank [270] and the EBI's Ensembl [132]. Via its Web API [165] the DDBJ provides Web Services for a variety of tasks. At the time of writing this book, 18 services (with a total of 124 individual operations) are available. Of these, the following are used in the examples presented in the following:

- ARSA is a DDBJ-specific keyword search system that supports simple and complex searches against a number of databases.
- GetEntry can be used to retrieve entries from a variety of common databases.
- Blast [29] searches databases for similar sequences.
- ClustalW (cf. Section 3.1.3) computes multiple sequence alignments.

- **Mafft** [150] is a multiple sequence alignment algorithm with different characteristics that provides an alternative to calling ClustalW.
- **PhylogeneticTree** creates JPG images from phylogenetic tree data.

These services provide quite common functionality in the sense that equivalent or similar services are also offered by other service providers. Most of the other services of the Web API are more DDBJ-specific.

3.2.2   Basic Multiple Sequence Alignment Workflow

In the simplest case, an alignment workflow in Bio-jETI consists of a SIB that calls an alignment service, and some SIBs that take care of handling the input and output data. Figure 3.6 shows such a basic alignment workflow: the first SIB (at the left, with the underlined name) lets the user select a file from the local file system, its content is then read into the execution context, the data is sent to the DDBJ's ClustalW web service for alignment computation, and finally the result is displayed.

Fig. 3.6 Basic alignment workflow, using the DDBJ's ClustalW web service

If an error occurs during execution of any service, an error message is written to the execution context by the respective SIB. This message can simply be displayed by the **ShowMessageDialog** SIB, or handled specifically, for instance by modeling fault tolerance or graceful termination of the computation with the possibility of resuming the workflow execution at the failed step later on. Note that in the example workflows that are shown here, all error handling is delegated to the calling workflow (by defining the **error** branches of the SIBs as model branches) to facilitate reading.

3.2.3   Variations of the Basic Workflow

Figure 3.7 shows a Bio-jETI workflow canvas with several preconfigured (i.e., with parameters already set) workflow snippets. In detail, they provide the following functionality:

1. Selecting and reading a sequence file from the local file system.
2. Calling the DDBJ's **ClustalW** alignment service.
3. Showing an alignment in a simple text dialog window.
4. Calling the DDBJ's **Mafft** alignment service.

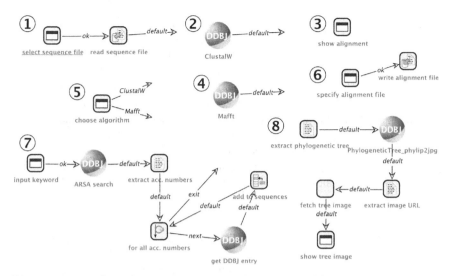

Fig. 3.7 Preconfigured parts for sequence alignment workflows

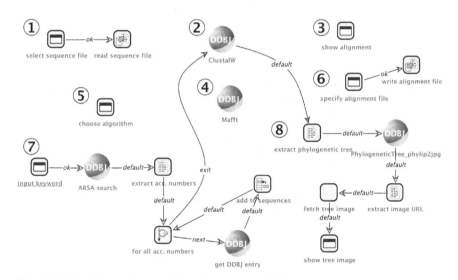

Fig. 3.8 Possible variation of the alignment workflow

5. Letting the user choose the service. If it is not known at workflow modeling time which alignment service shall be used, it is natural to leave the choice to the user. For this purpose, an interactive SIB can be used that displays a customized message dialog at runtime, asking the user to make a decision. In this case the user can choose between ClustalW and Mafft. According to which service has been chosen the SIB directs the flow of control.

6. Saving the alignment to the local file system. Storing results of analyses permanently rather than (only) showing them to the user at one point is useful whenever they are relevant outside of the specific workflow.

7. Letting the user enter a keyword, which is then used for a DDBJ database search (via the ARSA system) that results in a list of accession numbers for which the corresponding sequences are fetched from the DDBJ database. As the input data for an alignment is usually not initially available at the local file system, this is an useful option for obtaining a set of sequences with particular characteristics.

8. Extracting the phylogenetic tree that is part of a ClustalW result (using a regular expression) and calling the `phylip2jpg` service of the DDBJ that converts the textual tree representation into an image, followed by retrieving and displaying the image.

These snippets can now be put together to various alignment workflows simply by connecting them with the appropriate branches. For instance, connecting the snippets 1, 2 and 3 results in the basic alignment workflow from above. Connecting the snippets 7, 2 and 8 forms a more complex workflow (depicted in Figure 3.8), comprising database search by keyword, sequence retrieval, alignment computation, and visualization of the implied phylogenetic tree.

Note that the examples shown here are really only some simple examples of multiple sequence alignment workflows, suitable to give a first impression of workflow modeling in Bio-jETI. They do not cover all the modeling capabilities of the system. The features shown in the different versions of the example can be combined, and the workflows can be extended with whatever functionality is available and encapsulated in a SIB. For instance, the DDBJ's BLAST web service can be used to retrieve a set of sequences based on similarity to a query sequence rather than a query keyword as in the example above. A corresponding workflow is among the DDBJ's workflow examples [4], which list a number of workflows that make use of the DDBJ services. And naturally also other institutes' data retrieval and sequence analysis services can be applied, for instance those provided by the EBI [250, 167, 118] or BiBiServ [127].

3.3   Constraint-Driven Design of Phylogenetic Analysis Workflows

This section describes and discusses a scenario for the loose programming of phylogenetic analysis workflows based on a large and complex domain model that comprises similar, but much more services than those described in the previous section. This comprehensive example is particularly suitable to demonstrate the advantages of (semi-) automatic, semantics-driven workflow composition.

The domain model is based on the sequence analysis services from the European Molecular Biology Open Software Suite (EMBOSS) [264]. EMBOSS is a rich collection of freely available tools for the molecular biology user community. It contains a number of small and large programs for a wide range of tasks, such as sequence alignment, nucleotide sequence pattern analysis, and codon usage analysis as well as the preparation of data for presentation and publication. At the time of this writing, the latest version of EMBOSS (Release 6.4.0 from July 2011) consists of around 430 tools, some derived from originally standalone packages. EMBOSS provides a common technical interface for the diverse tools that are contained in the suite. They can, for instance, be run from the command line, or accessed from other programs. What is more, EMBOSS automatically copes with data in a variety of formats, even allowing for transparent retrieval of sequence data from the web. The tools work seamlessly for a number of different formats and types, and therefore the domain modeling can focus on the actual service semantics rather than on technical details of data compatibility.

The interfaces of the EMBOSS tools are described in so-called *ACD* (Ajax Command Definition) files, which provide information about the inputs and parameters that are required for the tool invocation. What is more, the ACD files contain annotations for tools, data and parameters that refer to terms from the EMBRACE Data and Methods Ontology (EDAM, cf. next section)[1]. This comprehensive meta-information about the EMBOSS tools facilitates the automated setup of a PROPHETS domain model for EMBOSS services (cf. [179]).

3.3.1   EMBRACE Data And Methods Ontology (EDAM)

The EDAM ontology [249, 138, 6] has been developed in the scope of the EMBRACE (European Model for Bioinformatics Research and Community Education) project [5] as an ontology for describing bioinformatics databases, data and tools that are in use in the community. It aims at providing a controlled vocabulary for the diverse services and resources in the Life Science Semantic Web that is suitable for bridging the gap between mere service registries and semantically aware service composition methodologies. Therefore it collects and combines knowledge from various existing resources, for instance from different web service collections such as the BioMoby registry [341] and the BioCatalogue [112], and from the myGrid ontology [347].

Importantly, EDAM is not a catalog of concrete services, data, resources etc., but a provider of terms for the annotation of such entities. Strictly

---

[1] At present, some ACD files contain incorrect, imprecise or inconsistent annotations, and also the EDAM ontology still contains some erroneous relations. As a consequence, some principally possible workflows are not recognized by the synthesis, and, conversely, some of the returned solutions are actually not possible. However, as both EMBOSS and EDAM are under active development, these issues will most certainly be solved within the next releases.

speaking, EDAM is not a single, large ontology, but consists of five separate
(sub-) ontologies:

- *Topic:* fields of bioinformatics study, such as *Phylogenetics*, *Polymerase chain reaction*, or *Metabolic pathways*.
- *Operation:* particular functions of tools or services (e.g., web service operations), such as *Phylogenetic tree construction*, *PCR primer design*, or *Sequence analysis*.
- *Data:* semantic descriptions of data entities that are commonly used in bioinformatics, such as *Phylogenetic tree*, *Sequence record* or *Sequence alignment*.
- *Identifier:* labels for the (unique) identification of entities, such as *Phylogenetic tree ID*, *Sequence accession*, or *Pathway or network name*.
- *Format:* references to (syntactic) data format specifications, such as *Phylip tree format*, *FASTA sequence format*, or *KEGG PATHWAY entry format*.

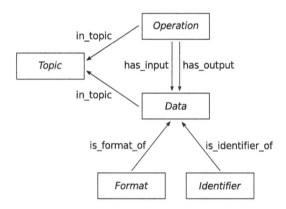

Fig. 3.9 Conceptual model of the EDAM ontology (following [6])

While the terms within the five (sub-) ontologies are simply organized
in subclass hierarchies, additional relations have been defined that express
further connections between these concepts, as Figure 3.9 shows. The *Format* and *Identifier* terms are linked to *Data* terms via *is_format_of* and
is_identifier_of relations, respectively. The *has_input* and *has_output* relations
can be used to denote the *Data* types that are consumed or produced by *Operations*. Finally, the *Operation* and *Data* terms can be associated to *Topics*
using the *in_topic* relation.

At present, EDAM contains more than 3000 terms and definitions. It is
the most comprehensive ontology of bioinformatics terms that is presently
available, and is becoming used by an increasing number of applications. As
detailed in [249], the major applications of EDAM envisaged by its developers

are the use of the defined terms for semantic annotations of web services in order to facilitate service discovery and integration, and the more detailed description of the involved data types in order to improve the data exchange between services.

In fact, a number of service providers (for instance the BioCatalogue [112] and BibiServ [127]) have started to annotate their services in terms of EDAM, and the latest release (6.4.0 of July 2011) of the European Molecular Biology Open Software Suite (EMBOSS) [264] contains EDAM annotations for the more than 400 tools of the suite and their parameters.

3.3.2   Domain Model

As any PROPHETS domain model, the domain model for this scenario consists of type and service taxonomies, semantically annotated services, and domain-specific workflow constraints. The domain-specific vocabulary that is provided by the taxonomies is used by both service annotations and constraints for referring to the types and services in the domain. Accordingly, the following introduces the type and service taxonomies that are used, before dealing with service annotation and constraint formulation.

Taxonomies

The taxonomies for this example are derived from the EDAM ontology described above, which provides terms for the classification of bioinformatics services, data and resources. The examples presented in this book are based on the beta12 version of EDAM, which has been released in June 2011 and is the EDAM version that has been used for annotating the EMBOSS 6.4.0 release on which this scenario is based. EDAM can simply be used for the service and type taxonomies by taking the (relevant parts of the) ontology and sorting the domain-specific service and data types into the skeletal domain structure. As detailed in [179], setting up a PROPHETS domain model based on EDAM involves four major steps:

1. Converting EDAM from OBO (Open Biomedical Ontologies) [285] into OWL [276] format.
2. Generating the service taxonomy from the *Operation* term and (transitively) all its subclasses.
3. Generating the type taxonomy from the *Data* and *Identifier* terms and (transitively) all their subclasses.
4. Sorting the available services and their input/output types into the service and type taxonomy, respectively.

Steps 1–3 can be executed fully automatic. Step 4 can be automated whenever EDAM annotations are available for the types and services, which is the case for the EMBOSS tool suite that is used in this scenario.

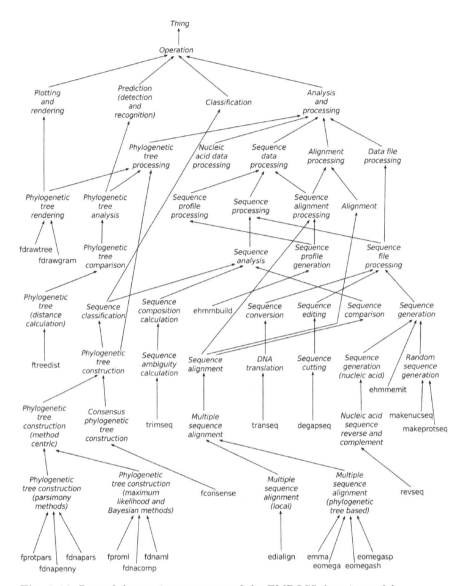

Fig. 3.10 Part of the service taxonomy of the EMBOSS domain model

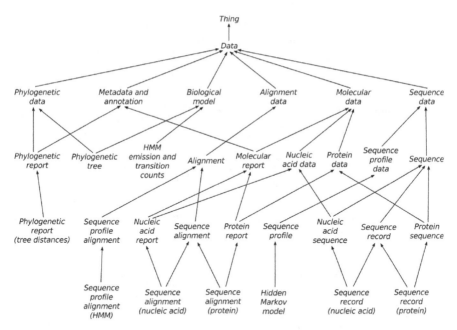

Fig. 3.11 Part of the type taxonomy of the EMBOSS domain model

Figures 3.10 and 3.11 show excerpts from the service and type taxonomies, respectively. The OWL class *Thing* is always the root of the taxonomies, below which EDAM terms provide groups for concrete and abstract service and type representations. Note that the taxonomies comprise 565 and 1425 terms, respectively, directly after being derived from EDAM. They are then automatically cut down to those parts that are relevant for the services and data that appear in the domain model in order to avoid overhead, still covering 236 and 207 terms, respectively. To facilitate presentation, the figures again comprise only those parts of the taxonomies that are relevant for the examples given in this section.

The (part of the) service taxonomy shown in Figure 3.10 comprises a number of service categories for different *Operation*s. The terms range from quite general service classifications (such as *Plotting and rendering* or *Analysis and processing* in the upper part of the figure) to rather specific service categories like *Phylogenetic tree construction (parsimony methods)* or *Multiple sequence alignment (local)* (lower part of the figure). The individual services in the domain model are then associated with one or more of the available service categories.

The (part of the) type taxonomy in Figure 3.11 contains various terms from the *Data* branch of the EDAM ontology. The higher-level data type categories comprise general terms like *Metadata and annotations* and *Phylogenetic data*, while the more specific categories at the lower levels allow for more precise

Table 3.1 Selection of services from the EMBOSS domain model

Service	Input types	Output types
degapseq	*Sequence record*	*Sequence record*
edialign	*Sequence record*	*Sequence alignment,* *Sequence record*
ehmmbuild	*Sequence record (protein)*	*Hidden Markov Model,* *Sequence alignment (protein)*
ehmmemit	*Hidden Markov Model*	*Sequence record (protein)*
emma	*Sequence record*	*Phylogenetic tree,* *Sequence record*
eomega	*Sequence record*	*Phylogenetic tree,* *Sequence record*
eomegash	*Sequence record,* *Sequence-profile* *alignment (HMM)*	*Phylogenetic tree,* *Sequence record*
eomegasp	*Sequence record,* *Sequence-profile*	*Phylogenetic tree,* *Sequence record,* *Sequence distance matrix*
fconsense	*Phylogenetic tree*	*Phylogenetic tree*
fdnacomp	*Sequence record* *(nucleic acid)*	*Phylogenetic tree*
fdnaml	*Sequence alignment* *(nucleic acid)*	*Phylogenetic tree*
fdnapars	*Sequence alignment* *(nucleic acid)*	*Phylogenetic tree*
fdnapenny	*Sequence alignment* *(nucleic acid)*	*Phylogenetic tree*
fdrawgram	*Phylogenetic tree*	*Phylogenetic tree*
fdrawtree	*Phylogenetic tree*	*Phylogenetic tree*
fproml	*Sequence alignment* *(protein)*	*Phylogenetic tree*
fprotpars	*Sequence alignment* *(protein)*	*Phylogenetic tree*
ftreedist	*Phylogenetic tree*	*Phylogenetic report* *(tree distances)*
makenucseq	-	*Sequence record*
makeprotseq	-	*Sequence record (protein)*
revseq	*Sequence record*	*Sequence record (nucleic acid)*
transeq	*Sequence record*	*Sequence record (protein)*
trimseq	*Sequence record*	*Sequence record*

data type classifications using terms like *Sequence alignment (nucleic acid)* or *Sequence record (protein)*.

Services

Table 3.1 lists the services that are relevant for the following examples, along with their input and output data types. The set of input types contains all mandatory inputs (i.e., optional inputs are not considered), while the set of output types contains all possible outputs. Note that the service interface definitions only consider the data that is actually passed between the individual services, that is, input parameters that are merely used for configuration purposes are not regarded as service inputs. The table comprises only 23 of the more than 430 services in the complete domain model. They provide functionality such as for the creation of molecular sequences (makenucseq, makeprotseq and ehmmemit), for basic processing of sequence data (e.g. trimseq and transeq), for phylogenetic analyses like alignments and phylogenetic tree construction (e.g. emma, fdnacomp), and for phylogenetic tree visualization (fdrawtree, fdrawgram).

Constraints

Initially, no domain constraints were defined for the EMBOSS domain model in order to maintain its full potential for experimentation. Later, some of the constraints that arose from the experimentation with the domain model that is described in the following were applied as domain-wide constraints. As the EMBOSS services constitute a really multi-purpose domain model (especially in contrast to the scenarios that are discussed in the next chapters), problem-specific constraints that are defined at workflow design time are more likely to be used.

3.3.3 Exemplary Workflow Composition Problem

When developing bioinformatics analysis workflows, users often have a clear idea about the inputs and final results, while their conception of the process that actually produces the desired outputs is only vague. Figure 3.12 (top) shows a simple loosely specified phylogenetics analysis workflow that reflects this starting point of workflow design: it begins with generating a set of random nucleotide sequences (using the EMBOSS service makenucseq) and ends with drawing and displaying a tree image (using fdrawtree and the viewer SIB of the jETI plugin), respectively. The first two SIBs are connected by a loosely specified branch (colored red). Note that the makenucseq service is used at this stage of the workflow design only to express the frame conditions in a convenient fashion: before the developed workflow would finally be released, this SIB would be replaced by a service that reads a meaningful nucleotide sequence from, for instance, a database or a file. The synthesis

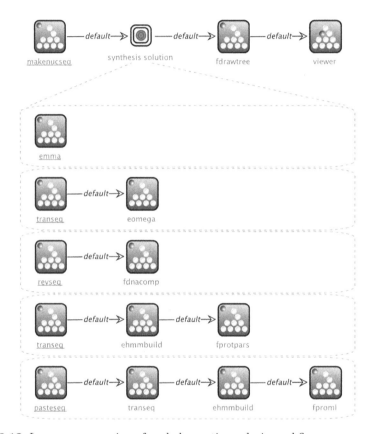

Fig. 3.12 Loose programming of a phylogenetic analysis workflow

problem that is defined by this loose specification is to find a sequence of services that takes `makenucseq`'s output (a nucleotide sequence) as input, and produces `fdrawtree`'s input (a phylogenetic tree) as output.

Applying the simplest synthesis process (simply returning the first solution that is found) to this specification results in the first of the possible concretizations that are shown in the lower part of Figure 3.12: A single call to `emma` (an interface to ClustalW), which produces a phylogenetic tree in addition to a multiple sequence alignment, solves the synthesis problem. However, there are also reasonable solutions to the synthesis problem that do not only contain a single phylogenetic tree construction service, but furthermore comprise varying numbers of, for instance, sequence editing, reformatting or preprocessing steps that define alternative analysis workflows for the same input/output specification.

The lower part of the figure shows four more examples of concretizations that may result from workflow synthesis for the loosely specified branch shown above if longer solutions are also considered: The second example consists of a call to `transeq` (translating the input nucleotide sequence into a protein sequence) followed by a call to `eomega` (an interface to the ClustalO protein alignment algorithm). In the third example, the reverse complement of the input sequence is built (`revseq`) and then used for phylogenetic tree construction with `fdnacomp`. In the the the fourth example the sequences are translated into protein sequences (`transeq`), which are then aligned via `ehmmbuild` and used for phylogenetic tree estimation with `fprotpars`. The last example solution is a similar four-step sequence where an additional sequence is pasted into the input sequences (`pasteseq`) and where `fproml` is used instead of `fprotpars` for the tree construction. Since EMBOSS provides various tools for phylogenetic tree construction as well as for the different sequence processing tasks, the solutions contained in the figure are by far not the only possible ones.

Accordingly, it is desirable to let the synthesis return further solutions in order to explore the possibilities that the domain model provides. When letting the synthesis perform a "naive" search in the synthesis universe (i.e., searching for all possible solutions considering only the input/output specification defined by the loosely specified branch) however, the algorithm encounters more than 1,000,000 results (the default limit of the search) for the synthesis problem already in search depth 4. While it is in principle possible to increase the limit and let the algorithm proceed to greater search depths and find further solutions, such a large number of solutions is not manageable for the user anyway. Moreover, although millions of solutions are easily *possible* with the described domain model, they are not necessarily *desired* or *adequate*. Hence, it is desirable to influence the synthesis process so that it returns less, but more adequate solutions.

The next section demonstrates in greater detail how "playing" with synthesis configurations and constraints helps mastering the enormous workflow potential. Therefore, it describes a simple but effective solution refinement strategy for the EMBOSS domain and the example workflow composition problem that drastically constrains the solution space by excluding clearly inadequate solutions and explicitly including adequate solutions.

3.3.4   Solution Refinement

Similar to the scenario that has been described in [180], this section describes a possible solution refinement strategy for the synthesis results that are returned by PROPHETS. It is of course only one highly specific example, tailored to a particular workflow composition scenario. A different use case will also require a different refinement strategy and finally a different set of domain-specific and problem-specific constraints. However, starting the solution refinement in a small search depth with constraints that exclude the most dispensable constructs, and then proceeding to greater search depths with constraints that include desired constructs explicitly, as it is done here, has turned out to be a common and expedient constraint development approach. Thus, the incremental development of the constraints is discussed at this level of detail only for this example, while for the other application scenarios simply the finally defined constraints are presented.

Table 3.2 Solution refinement for the exemplary workflow composition problem, considering solutions up to length 4

Refinements	Number of solutions			
	depth 1	depth 2	depth 3	depth 4
none	2	427	46,992	¿ 1,000,000
1	2	193	10,130	385,518
1, 2a	2	169	7,933	274,664
1, 2a, 2b	0	1	132	8,801
1, 2a, 2b, 3a	0	0	2	271
1, 2a, 2b, 3a, 3b	0	0	1	134
1, 2a, 2b, 3a, 3b, 4	0	0	1	34

Table 3.2 surveys the numbers of solutions for the different search depths and refinements steps, clearly showing the impact of the individual constraints on the solution sets. The first row of the table shows that when no constraints are applied, the synthesis algorithm finds two one-step solutions that solve the workflow composition problem, 427 solutions in depth 2, almost 47,000 solutions in depth 3 and exceeds the default limit of 1,000,000 solutions in depth 4. The remaining rows document how the incremental refinements that are described in the following successively narrow the search to the actually intended solutions.

In addition to the numbers given in Table 3.2, Figure 3.13 illustrates the impact of the constraints on the solution space graphically: It shows the

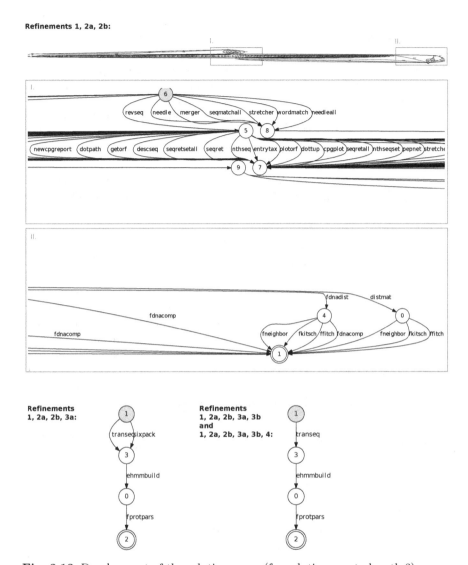

Fig. 3.13 Development of the solution space (for solutions up to length 3)

solution graphs, that is, compact result representations that aggregate all solutions in a single graph structure, for the third, fourth and fifth refinement steps when applied to the search for solutions up to length 3. In its entirety representative also for other application scenarios, this series of solution graphs illustrates the considerable impact of expressing concrete intents via constraints.

Due to the thousands of solutions that are obtained for the unconstrained case and the first two refinement steps, the corresponding solution graphs are extremely large and broad and therefore not reasonably representable on normal paper page size. Hence, the first solution graph shown in the figure is the graph that represents the 132 solutions that are obtained when refinements 1, 2a and 2b are applied. As it is still extremely broad, is it shown as a whole at the top of the figure, and additionally its two particularly interesting parts are shown somewhat enlarged below. The first one comprises the start state (shaded in gray) and some of the states in search depths 1 and 2. It is clearly visible that there are already quite a number of possibilities for the first step from the start state, and then extremely many for the second. As visible in the second enlarged part, of the graph, which comprises an accepting state of the solution graph in depth 3, less possibilities remain for the third step. The graphs for the solution sets for the last three refinement steps, shown at the bottom of the figure, are then indeed extremely small, only representing two or one solutions.

Refinement 1: Exclusion of Useless Services

When taking a look at the actual solutions, it is striking that several of the solutions that are obtained when no constraints are applied contain services that are useless in the sense that they distract from the actual synthesis problem. Such services are, for instance, services that require no input but produce new data that is planted into the workflow, like the `makenucseq` and `makeprotseq` services that create sequence data "from scratch". To avoid the inclusion of such services in the solutions, a constraint can be defined that excludes all services that do not require any inputs and thus may create superfluous data. Therefore, the service ontology has been extended by an auxiliary *ServiceWithNoInput* class, under which all these services were grouped, and then the *exclude service* constraint (cf. Section 2.3.2) could simply be used:

• Do not use *ServiceWithNoInput*.

As Table 3.2 (row 2) shows, this constraint decreases the number of solutions that are found considerably: the two one-step solutions remain, but only 193 solutions are found in depth 2, then 10,130 in depth 3, and 385,518 in depth 4.

Refinement 2: Exclusion of Unwanted Services

Another observation about the solutions is that some of the workflows contain services that do in fact perform operations on the available data, but which are actually unwanted within the intended solution. For the present example, for instance, it can be assumed that the input sequence data is suitable for further analysis and does not have to be edited beforehand. Accordingly, *Sequence editing* services (like, e.g., `degapseq`) can explicitly be excluded from the solutions, again using the *exclude service* constraint:

• Do not use *Sequence editing*.

This constraint decreases the number of solutions further down to 169 in depth 2, then 7,933 in depth 3, and 274,664 in depth 4 (see row 3 of Table 3.2).

Similarly, it can be observed that often multiple sequence alignment services (like, e.g., *emma*) are used, which produce phylogenetic trees as more ore less by-products of the alignment computation. If it is not desired to derive phylogenetic trees this way, also *Sequence alignment* services can be excluded from the solutions:

- Do not use *Sequence alignment.*

With this constraint, no one-step solutions are found any more and there is only a single two-step solution (see row 4 of Table 3.2). Furthermore, the number of solutions found until a depth of 3 is decreased further to 132, and to 8,801 until a depth of 4.

Refinement 3: Expression of Intents

Still, this number of solutions is not really manageable for the user. In order to steer the search towards the intended solutions more efficiently, the synthesis algorithm can be provided with constraints that express specific ideas about the desired workflow.

For instance, the *service enforcement* constraint can be used to explicitly demand the inclusion of a phylogenetic tree construction service that uses a parsimony method:

- Enforce the use of *Phylogenetic tree construction (parsimony methods).*

With this constraint, first solutions are found in depth 3, and the synthesis returns a total of 271 solutions in depth 4 (see row 5 of Table 3.2).

If the user even has concrete tools in mind that he would like to see in the solution, he can enforce their existence. For instance, if the input nucleotide sequence should be translated into a protein sequence by `transeq` and then a phylogenetic tree be created using `fprotpars`, corresponding constraints can be applied:

- Enforce the use of `transeq`.
- Enforce the use of `fprotpars`.

These constraints narrows the search to a single solution of length 3 and a still assessable set of 134 solutions in depth 4 (see row 6 of Table 3.2).

Refinement 4: Alternative Synthesis Strategy

Another, general observation about the obtained solution is that often workflow parts without proper relationships to the surrounding workflow are created. Most prominently, outputs of services are often not adequately used in the subsequent workflow. However, a strict output-input chaining, guaranteeing that the data created by a service is used only by its immediate successor, is a property that can not generally be expressed in terms of SLTL constraints.

Interestingly, PROPHETS provides an alternative to the default synthesis processes (which assumes a shared memory for exchanging data), where strict pipelining is mimicked and service outputs are only transferred to direct successors and are thus not available for any other subsequent services. Applying this alternative synthesis strategy reduces the number of solutions to be considered significantly, as the last row of Table 3.2 shows, leaving an easily manageable set of only 34 solutions.

For domains like the EMBOSS example, which consist of a large number of services which furthermore have mostly only one primary input, this search strategy can in fact be useful for constraining large sets of solutions further. One should be aware, however, that using the pipelining synthesis process of course leads to missing some solutions, namely those that involve services with several inputs which come from different preceding services. Thus, this synthesis process is in general not adequate and has not been considered further for the applications discussed in this book.

Greater Search Depths

Having identified constraints that reduce the search space to a size at which a manageable set of solutions is obtained, it might be interesting to consider greater search depths again. Other, interesting solutions might be found that perform analyses that have not been seen so far, as they are not possible with shorter service sequences. For instance, applying all constraints developed above and proceeding until a search depth of 5, the synthesis finds 458 additional solutions. These solutions are mostly new combinations of the services that have already been used within the shorter solutions, but some solutions also make use of services that have not been contained earlier.

GeneFisher-P

This second application scenario is based on the GeneFisher web application for PCR primer design [109, 124]. To provide a more flexible alternative to the monolithic web application, it has been realized as Bio-jETI workflow. With the resulting GeneFisher-P [177], workflow variants that are not covered by the web application, such as using alternative services for individual analysis steps or batch processing of input data sets, can easily be built at the user level.

4.1 Background: PCR Primer Design

A key technology within the discipline of genetic engineering and one of the most widely used techniques in biological laboratories today is the polymerase chain reaction (PCR) [338], developed around 1985. It can be pictured as "a kind of genetic photocopying" [55, p. 291]: it allows for isolating and exponentially amplifying short fragments from a DNA sequence (up to around 6000 bp) without utilizing living cells for the reproduction. Analysis of genetic information is only possible when a sufficient amount of (identical) DNA is available, and so PCR is used in biological research projects (for instance for DNA sequencing and functional analysis of genes), and medicine (diagnosis of hereditary diseases, detection of genetic defects), as well as in forensic sciences and paternity testing (identification of genetic fingerprints).

4.1.1 Polymerase Chain Reaction

A PCR run consists of a series of cycles (usually 30-60) which are essentially carried out in three steps (see Figure 4.1):

1. The *denaturation step* melts the double-stranded DNA into two separate strands at a temperature of 94-96 °C. In the first cycle, this steps needs

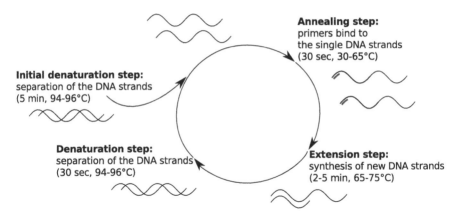

Fig. 4.1 PCR cycle

around 5 minutes, in the following cycles around 30 seconds are sufficient, since only the newly synthesized, short DNA fragments from the
preceding cycle have to be separated again.

2. During the *annealing step*, primers bind to the single-stranded template.
   Primers are short oligonucleotides that constitute the starting points for
   the DNA synthesis. For PCR, two (usually synthetic) primers are needed
   that bind correctly at the 5' and 3' ends of the DNA region of interest.
   The temperature that is applied in the annealing step depends on the
   melting temperature of the concrete primers.

3. In the *extension step* (2-5 minutes at 65-75 °C), a DNA polymerase syn-
   thesizes new DNA strands complementary to the single-stranded templates.

This procedure is then repeated, doubling at each step the number of
copies of the desired DNA segment. Through such repetitive cycles, which are
carried out fully automated by modern laboratory equipment, it is possible
to obtain millions of copies of single DNA segments within a short period
of time. Moreover, PCR is applicable to detect the presence of a specific
target DNA in a complete genome or even in a mixture of DNA from various
organisms: due to the high specificity of the PCR primers, they will only bind
and amplify a very limited set of DNA fragments.

4.1.2   PCR Primer Design

Before each PCR experiment, specific primers have to be designed, taking into
account a number of criteria. The method described in [109] and sketched in
Figure 4.2 uses a number of global primer properties (such as melting temperature, GC content, size, degeneracy, and uniqueness) and a number of
properties that refer to the 3' end of the primer (GC content, degeneracy,
terminal residue). The search for suitable primers for a input sequence s is

then straightforward (left side of the figure): s is run through and at each position the theoretically possible primers (from minimum to maximum primer length) are tested against the given global and 3' end properties. If a subsequence of s passes all tests, it is added to the pool of possible (forward) primers. Then, s is reversed and complemented and the procedure is repeated in order to obtain a pool of primers for the reverse direction as well. Finally (right side of the figure), primer pairs are built by matching each primer from the forward pool to each primer from the backward pool. Only those pairs are kept that conform to the given criteria regarding the size of the amplified region, the avoidance of dimer structures (such as hairpin loop structures and primer-primer interactions) and melting temperature compatibility.

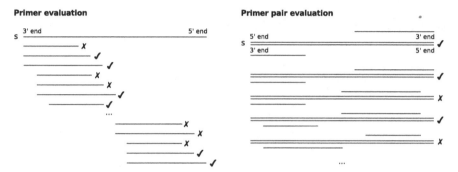

Fig. 4.2 Primer design for a sequence s

As mentioned in the previous section, the PCR method can also be used for detecting particular DNA segments in mixtures of DNA from various organisms. Suitable degenerate primers can be designed from sequences of homologous genes, based on the assumption that genes with related function from different organisms have a high sequence similarity. In this case, the primer design process starts with the computation of a multiple sequence alignment from the homologous sequences. A consensus sequence is then calculated from the alignment and used for the primer design as described above.

4.1.3 GeneFisher

The GeneFisher software [109] implements the method for designing degenerate primers described above. The term "gene fishing" refers to the technique of using PCR to isolate a postulated but unknown target sequence from a pool of DNA.

GeneFisher accepts single or multiple DNA and protein sequences as input. As primers are always calculated for a single DNA sequence, multiple input sequences are aligned using alignment programs such as ClustalW (cf. Section 3.1.3), DCA [312] or DIALIGN [226]. From the alignment, a consensus sequence is derived and used as input for the primer calculation step. If protein sequences have been submitted, a backtranslation step is also necessary, where amino acid sequences are translated to a hypothetical nucleotide sequence using either maximum redundancy methods or the codon usage of a given organism.

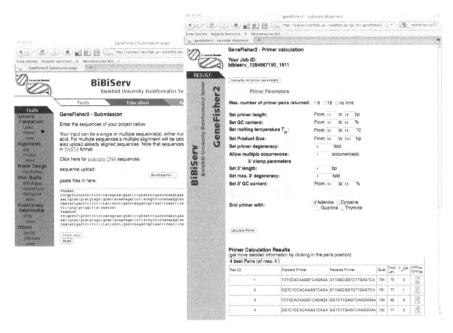

Fig. 4.3 GeneFisher2 web interface

GeneFisher has been hosted as a web application at the Bielefeld Bioinformatics Server (BiBiServ, [127]) for several years. Figure 4.3 shows the web interface of its recent reimplementation, GeneFisher2 [124]: The first step in a GeneFisher session is to provide the input sequence(s), either via file upload or by pasting the data directly into a text field (left side of the figure). After sequence submission the user chooses an alignment tool and the parameters. Before consensus generation the alignment can be inspected and re-run with adjustments, if necessary (not shown in the figure). Based on the consensus sequence and the primer parameters, the primers are calculated (right side of the figure). If GeneFisher can not calculate any primers for given parameters, a rejection statistics suggests which parameters the user should change (not shown).

4.2 GeneFisher-P Workflows

GeneFisher-P [177] is an example of a more complex workflow: PCR primer design with the GeneFisher application comprises an alignment step in case the input consists of multiple nucleic or amino acid sequences. Depending on the type of input, services for sequence alignment computation, consensus calculation, backtranslation, and finally for the primer design itself are invoked by the process. Figure 4.4 summarizes how the different kinds of inputs are processed before the actual primer calculation based on a single nucleotide sequence can take place: Multiple input sequences have to be aligned and transformed into a (single) consensus sequence, protein sequences have to be translated (back) into an equivalent nucleotide sequence.

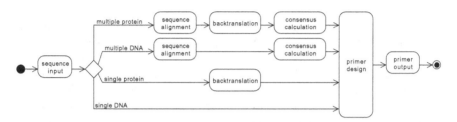

Fig. 4.4 Abstract GeneFisher workflow (following [124])

GeneFisher is a completely predefined application. Its internal workflow is hidden from the end user, who can only interact with it through the web GUI. The limitations of this interface become obvious in particular when it comes to the full automation of the primer design process (batch processing) or the exchange of single components. The objective behind GeneFisher-P (the *-P* denotes the process-based realization of the application) was to put the processes in the foreground and expose the internal workflows and the underlying services and components to the end user, who becomes then able to intervene and change or integrate them with others according to his specific needs (examples are given in Section 4.2.4).

4.2.1   GeneFisher Services

The tools that are used by the GeneFisher web application and likewise for its GeneFisher-P workflow realization, such as the alignment programs ClustalW and DCA, are all hosted at BiBiServ. Some of them provide web service interfaces, others are legacy programs that have been integrated using the jETI technology (cf. Section 2.1.2). These tools originate from the first days of the GeneFisher project, have been written in C and compiled for a particular CPU type, and are thus dependent on specific machines, which means that they can cannot be made available via the internet directly. The following services are central within GeneFisher-P:

- ClustalW (cf. Section 3.1.3), which is available as web service at BiBiServ.

- DCA [312] stands for *Divide-and-Conquer Multiple Sequence Alignment*. The program is very fast and produces high-quality multiple sequence alignments of protein, RNA, or DNA sequences. It is also provided as a web service at BiBiServ.
- BatCons is one of the mentioned legacy programs from the original Gene-Fisher project. It performs backtranslation (in case of protein sequences) and consensus calculation (if the input consists of multiple sequences).
- gf_2000, another legacy program, is responsible for the actual primer design. The input has to be a single nucleotide sequence.

Local tasks like input validation, user interaction and the definition of the control flow, which are programmed in the case of GeneFisher, are modeled in GeneFisher-P by jABC Common-SIBs that encapsulate corresponding functionality. For the control flow, these are mainly SIBs declaring variables or checking conditions, as well as for steering the flow of control in dependence of intermediate results. User interaction refers, for instance, to SIBs for file management and data input, as well as for display of (intermediate) results. In the GeneFisher web applications, the user can choose between ClustalW and DCA for the multiple alignment. For simplicity, the GeneFisher-P workflows are modeled using only the ClustalW service. However, analogously to what has already been shown for the previous scenario, ClustalW can easily be exchanged with DCA by the user.

4.2.2 Interactive GeneFisher-P Workflow

Different workflow realizations of the GeneFisher process as summarized in Figure 4.4 have been built in the scope of the GeneFisher-P project. Figure 4.5 shows a simple interactive version: The workflow starts with asking the user for a single input sequence file and a classification of the sequences (i.e., single or multiple sequence, nucleotide or amino acid). According to the sequence type, an appropriate primer design subworkflow is called. Finally, the result of the primer design is stored into a file specified by the user.

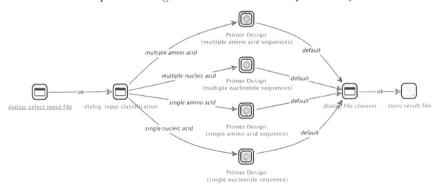

Fig. 4.5 Interactive GeneFisher-P workflow

Fig. 4.6 Primer design for multiple amino acid sequences

The subworkflows are tailored to a specific input sequence type, according to the abstract workflow given in Figure 4.4. Figure 4.6 shows the subworkflow for primer design based on multiple amino acid sequences: the given input file name is transformed into an URI (ensuring that the file is readable at each platform), its content is read, and then sequence alignment, backtranslation and consensus calculation are performed prior to the actual primer design. Finally, the result is written into a file. The subworkflow for primer design based on multiple nucleotide sequences is highly similar, only the backtranslation step is omitted. In case of a single amino acid sequence, backtranslation is necessary, but the alignment and consensus calculations steps are not required. Single nucleotide sequences need no preprocessing at all.

4.2.3   Batch Processing GeneFisher-P Workflow

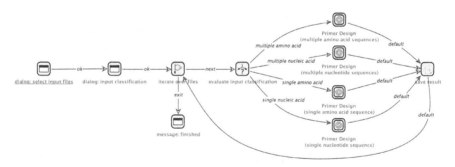

Fig. 4.7 Batch processing GeneFisher-P workflow

As mentioned before, it is not feasible to design large numbers of primers when the process has to be carried out interactively (which is the case for the very interactive GeneFisher web application as well as for the GeneFisher-P workflow discussed before, where basic information about the data has to be provided at runtime). The workflow shown in Figure 4.7 reduces the required interaction to a minimum: only the input files and the types of the input sequences have to be specified, then the files are worked off by the corresponding subworkflow within a loop autonomously in a batch processing manner.

4.2.4 *Further Variations of the GeneFisher-P Workflow*

The interactive and the batch processing workflows are only two examples of useful GeneFisher workflow realizations. The journal article on GeneFisher-P [177], for instance, describes further workflow variations, similar to the alignment workflow examples in Section 3.2 and also incorporating external services: alternative ClustalW implementations or other multiple sequence alignment methods are used (for instance via the EBI [250, 167, 118] or DDBJ [165] Web Services), or the workflows are extended by automated data retrieval, for instance by retrieving the input sequences from a remote database. In a variation that resembles the very interactive nature of the original GeneFisher web application closer than the workflow shown in Figure 4.5, also intermediate results are displayed to the user and may be accepted or rejected. In case of rejection, the workflow leads the control-flow back to a previous process step, typically to a point where the appropriate data or parameters can be modified before the corresponding part of the workflow is executed again.

At the beginning of this section it was stated that the objective behind GeneFisher-P was to create a GeneFisher version that could easily be batch-processed in order to design large numbers or primers, and where single components could be easily exchanged by services that meet the specific needs better. Both these features where in fact applied in a chelicerate phylogeny study at the University of Hamburg [50], where several pairs of PCR primers were designed using GeneFisher-P. The multiple alignment was carried out using Mafft [150] instead of ClustalW, and the backtranslation and consensus calculation via the legacy BatCons tool was replaced by a new Ruby script that implements backtranslation and consensus calculation with some specific features that cope better with the characteristics of the data that were used in the study.

4.3   Constraint-Driven Design of GeneFisher-P Workflows

This section shows how loose programming can be applied for the development of the GeneFisher-P workflows. As the primer design process is performed by a set of rather common services (alignment, backtranslation, consensus calculation, etc.) the EDAM terminology (cf. Section 3.3.1) can also be used for the annotation of the services in the GeneFisher-P domain model. However, the domain model for GeneFisher-P is much smaller than the one for EMBOSS, and hence less constraints are necessary for keeping the search and solutions within a feasible range.

4.3.1   Domain Model

The following describes the service and type taxonomies, service descriptions and domain constraints that have been defined for the GeneFisher-P domain model.

Taxonomies

Figure 4.8 shows the service taxonomy of the GeneFisher-P domain model. The OWL class *Thing* is the root of the taxonomy, below which the EDAM term *Operation* explicitly provides a group for concrete and abstract service representations. The taxonomy comprises a number of service categories for different operations concerned with the design, comparison, analysis and processing of sequence data. As the figure shows, the concrete services (available as SIBs) are then sorted into the taxonomy. The different file processing services are only quite generally classified as *Operation* (upper part of the figure). The actual GeneFisher services are then more specifically associated with the *DNA backtranslation, Sequence alignment conservation analysis, Multiple sequence alignment* and *PCR primer design* classes (lower part of the figure).

The type taxonomy of the GeneFisher-P domain model is shown in Figure 4.9. It contains terms from both the *Identifier* and *Data* branches of the EDAM ontology. While the former finally only contains the *File name*, the latter describes the *URI* data type more precisely, defines an explicit *Sequence alignment (multiple)* term and comprises a number of categories for different kinds of sequence data.

Services

Table 4.1 lists the services contained in the GeneFisher-P domain model (in alphabetical order) along with their input and output type annotations in terms of the EDAM ontology. The set of input types contains all mandatory inputs (i.e., optional inputs are not considered), while the set of output types contains all possible outputs. Note that the service interface definitions only consider the data that is actually passed between the individual services, that is, input parameters that are used for configuration purposes are not regarded as service inputs. As also visible from the service taxonomy (Figure 4.8), the GeneFisher-P domain model comprises the actual GeneFisher services for sequence processing, alignment computation and PCR primer design, as well as a number of services providing basic file processing functionality.

Constraints

Finally, a set of domain constraints is defined that covers general properties of GeneFisher-P workflows, such as that the `GeneFisher` service that actually designs the PCR primers is always included, that the individual analysis steps are not repeated unnecessarily, or that steps are executed in a particular order:

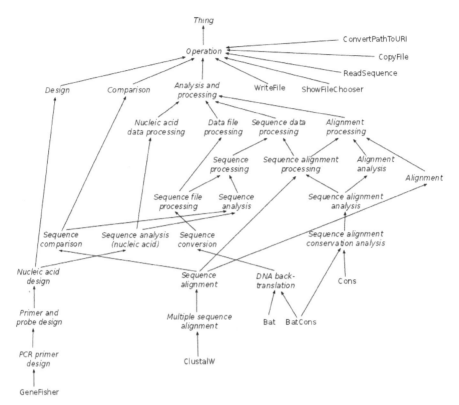

Fig. 4.8 Service taxonomy of the GeneFisher-P domain model

- Enforce the use of `GeneFisher`.
- Do not use *DNA backtranslation* more than once.
  Do not use *Sequence alignment conservation analysis* more than once.
  Do not use *Sequence alignment* more than once.
  Do not use *PCR primer design* more than once.
- If *PCR primer design* is used, do not use *Sequence alignment* subsequently.
  If *PCR primer design* is used, do not use *Sequence alignment conservation analysis* subsequently.
  If *PCR primer design* is used, do not use *DNA backtranslation* subsequently.
- Use service `WriteFile` as last service in the solution.

Additional, problem-specific constraints (defined in the workflow design phase) can narrow the possible solutions further, for instance by explicitly working with single/multiple nucleic acid/amino acid sequences and the accordingly required processing steps (cf. Section 4.2).

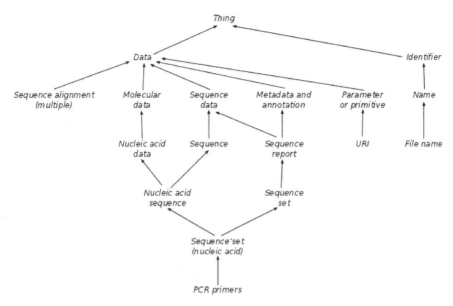

Fig. 4.9 Type taxonomy of the GeneFisher-P domain model

Table 4.1 Services in the GeneFisher-P domain model

Service	Input types	Output types
Bat	*Sequence*	*Nucleic acid sequence*
BatCons	*Sequence alignment (multiple)*	*Nucleic acid sequence*
ClustalW	*Sequence*	*Sequence alignment (multiple)*
Cons	*Sequence alignment (multiple)*	*Nucleic acid sequence*
ConvertPathToURI	*File name*	*URI*
CopyFile	*File name*	
GeneFisher	*Nucleic acid sequence*	*PCR primers*
ReadSequence	*URI*	*Sequence*
ShowFileChooser		*File name*
WriteFile	*Data*	*File name*

Loosely specified workflow:

Possible concretizations:

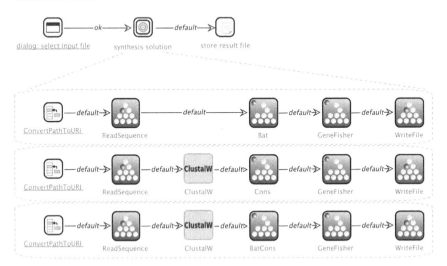

Fig. 4.10 Loose programming of a GeneFisher-P workflow

4.3.2   Exemplary Workflow Composition Problem

Figure 4.10 (top) shows an example of a loosely specified GeneFisher work-flow: Instead of modeling one of the (four) variants of the primer design workflow explicitly, only the starting point (selection of an input file) and the end point (storing of the result) are given. The connecting loose branch can then be concretized dynamically into the currently intended, concrete primer design workflow.

The lower part of the figure shows three possible workflow concretizations, corresponding to variants of the GeneFisher-P workflows as described in Section 4.2. The first variant simply performs the backtranslation and primer design that is suitable for an input file that contains a single amino acid sequence, while the third and fourth variant additionally perform the alignment computation, backtranslation and consensus calculation steps that have to be performed prior to primer design in case of an input file that contains multiple amino acid sequences. The inclusion of the respective services in the solution can simply be achieved by constraints that enforce their existence. Likewise, the other variants of the primer design workflow can be created by varying the constraints.

Considering all solutions until a search length of 6, the synthesis finds 650 possible workflows with the available services in the completely unconstrained case. Applying the domain constraints reduces the number of returned solutions to an easily accessible set of 8 solutions. Additional constraints that enforce the use of a *DNA backtranslation* and a *Sequence alignment* service (as required for an input consisting of multiple amino acid sequences) even lead to one single solution. In contrast, when avoiding the use of *DNA backtranslation* and *Sequence alignment* services (as required for single amino acid sequences), the specification is over-constrained and no solution is found, that is, the minimal GeneFisher-P workflow that directly calls the `GeneFisher` service with the input sequence is not recognized by the synthesis. The reason is that, according to the domain model, `GeneFisher` requires a *Nucleic acid sequence* as input, while `ReadSequence` produces a more general *Sequence*. In order to fix this problem, additional services with more concise interface descriptions would have to be introduced, for instance separate `ReadNucleotideSequence` and `ReadAminoAcidSequence` services that allow for a more fine-grained consideration of the involved data types.

5

FiatFlux-P

This third application scenario is concerned with the automation of time-consuming metabolic flux analysis procedures based on the software Fiat-Flux [353]. In addition to the integration of its principal functionality as elementary services, new services were implemented that emulate the user interaction with FiatFlux in order to enable automation. Like GeneFisher-P, the resulting *FiatFlux-P* facilitates building variants and defining batch processing workflows at the user level.

5.1   Background: Metabolic Flux Analysis

Rapidly improving analytical techniques in the life science laboratories, especially in the field of high-throughput procedures, have made it possible to investigate biological phenomena on a more comprehensive scale. It has become common practice to consider the entirety of genes, RNA, proteins or metabolites that are present in the cell at a particular time point, in addition to the examination of single molecular structures. These comprehensive datasets can be used to infer metabolic pathways, complete genetic regulatory or reaction networks and be integrated to obtain comprehensive (computational) models of biological systems [279, p. 171]. This approach, aiming at "understanding biological organisms in their entirety" [279, p. 273] via integration of data from genomics, transcriptomics, proteomics, metabolomics, and phenomics measurements into computational models, has been coined as *Systems Biology*.

5.1.1   *Metabolomics*

Metabolomics, one sub-discipline of Systems Biology, "deals with the analysis of metabolites" [279, p. 262]. More precisely, it aims at studying the quantitative occurrences of metabolites within metabolic pathways that are possibly already known qualitatively. There is currently no technique available that can measure the entirety of all cellular metabolites simultaneously, but

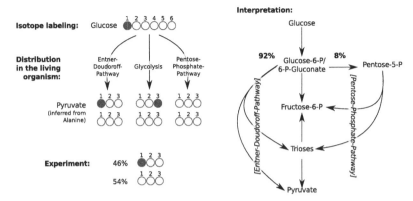

Fig. 5.1 Metabolic flux analysis based on ^{13}C labeling experiments

methods exists that can identify and quantify defined sets of compounds in parallel. The most frequently applied methods for metabolic profiling are *nuclear magnetic resonance* (NMR) and *mass spectrometry* (MS), where the latter is usually preceded by a chromatographic step (such as *gas chromatography*, GC, or *liquid chromatography*, LC) to separate the metabolites. Knowledge from metabolomics experiments can be used for building or refining models of metabolic pathways. As metabolites are at the end of the cellular information chain [279, p. 165 ff.], their quantification also allows to draw inferences about other cell networks, such as gene regulatory networks or signal transduction pathways.

Figure 5.1 sketches the procedure of a so-called ^{13}C *tracer experiment*: Glucose, a sugar consisting of six carbon atoms, is labeled with a ^{13}C isotope at the first carbon atom of the sugar molecule (upper left of the figure). The marked glucose is then fed to the biological system and, due to the catabolism of the sugar molecule, the labeled atoms are distributed over the metabolic network. Later on, the labeling patterns of intracellular metabolites or amino acids located in proteins can be measured by NMR or MS instruments. The resulting spectra provide information about the physiology of the biological system under investigation. Depending on the organism, there are several possibilities of how the cell processes the glucose into pyruvate (three C atoms), with the purpose of gaining energy from the molecule splitting. The microorganism Escherichia coli, for instance can metabolize glucose via the *Entner-Doudoroff pathway*, the *glycolysis*, and the *Pentose-Phosphate pathway*. As the figure (upper left) shows, the Entner-Doudoroff pathway results in two pyruvate molecules of which is unlabeled and the second contains the ^{13}C of the glucose at the first carbon position. Glucose catabolism via the glycolysis results in a 50:50 mixture of unlabeled pyruvate and molecules carrying the ^{13}C isotope at the C_3 position. When glucose is processed via the Pentose-phosphate pathway, neither of the resulting pyruvate molecules contains a ^{13}C atom. In the experiment, the labeled glucose is fed to an organism, and the resulting pyruvate molecules are measured and classified according to the scheme describe above (lower left). If, for instance, 54% of the pyruvate is unlabeled and 46% carries a ^{13}C at the first position, it can be inferred that 92% of the glucose was processed via the Entner-Doudoroff pathway, and 8% via the Pentose-phopshate pathway (right side of the figure).

5.1.2   Metabolic Flux Analysis

Flux distributions, that is, the set of reaction rates (= fluxes) of the system under consideration, are the integrated network responses of the different cell components (genes, mRNAs, proteins, and metabolites) and abiotic physico-chemical system parameters. They can be used to decipher cellular functions and guide rational strain engineering for industrial biotechnology. In contrast to the quantification of cellular components (mRNA, proteins and metabolites), fluxes are not directly accessible by experimental techniques but have to be inferred using mathematical models of the cellular metabolism.

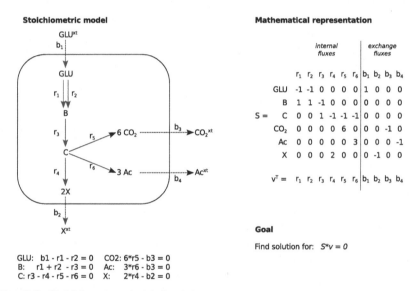

Fig. 5.2 Stoichiometry at steady state

Assuming that the system is in a stationary or quasi-stationary state (i.e., that metabolite concentrations and reaction rates do not change over time), reaction rates can be calculated applying metabolic flux analysis, which is based on metabolite mass balances. Figure 5.2 gives an example: Knowledge about the stoichiometry of the reaction network (composed of the the internal fluxes $r_1, \ldots r_6$ and extracellular uptake and secretion rates, the exchange fluxes $b_1, \ldots b_4$) is sufficient to set up a linear equation system. In matrix notation this linear equation system reads: $S * v = 0$. The matrix S contains the stoichiometric coefficients of the metabolites in the respective reactions (the rows of the matrix correspond to the metabolites, the columns correspond to the reactions). Together with the flux vector v, composed of the reaction rates of the unknown internal reactions rates r and the partially experimentally determined exchange reactions b, and the assumption of a steady state, the computational task remains to find a solution for $S * v = 0$. This task can be achieved by the tools that linear algebra provides (details would be beyond the scope of this book, the interested reader is referred to [230] for further information). For most reaction networks, the system is underdetermined if only constrained by extracellular uptake and secretion rates and the growth rate of the cell, meaning that often not all fluxes, especially those of parallel pathways and cycle fluxes of the network, can be resolved.

Additional constraints are gained from growth experiments with stable isotope tracers like ^{13}C (cf. Section 5.1.1). The data can be used to estimate the flux distribution inside the cell of interest. The rationale behind these ^{13}C tracer experiments is that the carbon backbones of the metabolites often are manipulated differently by alternative pathways, leading to different ^{13}C labeling patterns of the metabolites. Thus, constraints to fluxes complementary to the basic stoichiometric constraints can be derived by measuring the mass isotope distribution of metabolites, that is, the relative abundances of molecules only differing in the number of heavy isotopes, which render the system fully resolvable.

Currently, two main approaches exist for such interpretation of the determined ^{13}C labeling patterns and the inference of intracellular fluxes. In the *global isotopomer balancing approach* [275, 339, 345, 259], the problem of estimating metabolic fluxes from the isotopomer measurements are formulated as a nonlinear optimization problem, where candidate flux distributions are iteratively generated until they fit well enough to the experimental ^{13}C labeling patterns. The second method is *metabolic flux ratio analysis*, coined as *METAFoR* [269], which relies on the local interpretation of labeling data using probabilistic equations, which constrain the ratios of fluxes producing the same metabolite. The approach is mainly independent of the global flux distribution in the entire metabolic network [269, 43, 94] meaning that flux ratios can be calculated without knowing the uptake and production rates of external metabolites and the biomass composition of the cell. If enough independent flux ratios can be identified, it is possible to use them to constrain the metabolic network equation system and to calculate the full flux distribution of the network [95].

5.1.3   FiatFlux

FiatFlux [353], developed to facilitate ^{13}C-based metabolic flux analysis, is a MatLab-based software that provides a user interface and interactive workflow for the analysis of GC-MS-detected ^{13}C patterns of proteinogenic amino acids. It allows for the calculation of flux partitioning ratios (METAFoR analysis) and absolute intracellular fluxes by ^{13}C-based metabolic flux analysis (net flux computation) [95]. Currently, the analysis is restricted to GC-MS derived mass spectral data of proteinogenic amino acids from cultures grown on mixtures of uniformly labeled glucose (U-^{13}C-glucose), 1-^{13}C-glucose and naturally labeled glucose. The software reads mass spectral data from ^{13}C-labeling experiments that is provided in netCDF format (CDF: common data format) and saves it in a .ff-file (FF: FiatFlux) for internal use. The data can then directly be analyzed in the modules `ratio` and `netto`, in which the METAFoR analysis and net flux computation are implemented, respectively.

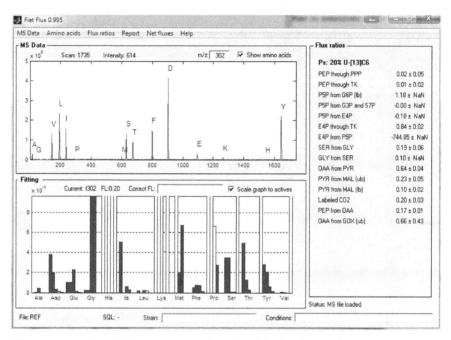

Fig. 5.3 FiatFlux desktop user interface for METAFoR analysis

Figure 5.3 shows the user interface of FiatFlux for the METAFoR analysis. When MS data is loaded, the software automatically detects and assigns the amino acid fragments (displayed in the "MS Data" frame). Additionally, the user has to specify the different experimental parameters. Then the mass distribution vectors of the amino acid precursors (MDV_M) are calculated by least square fitting (and displayed in the "Fitting" frame) and flux ratios are estimated using probabilistic equations (and displayed in the "Flux ratios" frame).

The visualization of the MDV_M facilitates the diagnosis of faulty measurements, such as outlier amino acid fragments that are indicated by abnormal residuals. To exclude such faulty or redundant data from the calculations, the user simply has to perform a single mouse click on the blue bar corresponding to a particular amino acid fragment, which is then colored white. Analogously, clicking on a white bar reactivates the corresponding fragment.

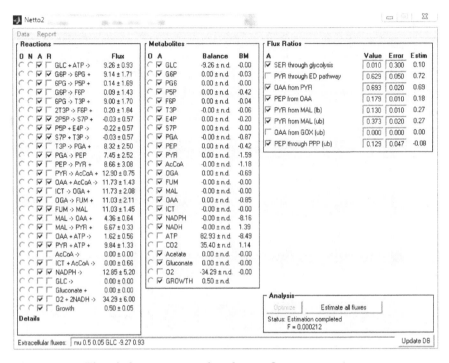

Fig. 5.4 FiatFlux desktop user interface for net flux computation

To compute net fluxes in the central carbon metabolism network, ^{13}C-based metabolic flux analysis is applied within the **netto** module of FiatFlux. Figure 5.4 shows its user interface, which allows for the detailed configuration of the reaction network model prior to net flux computation: In the "Reactions" frame (left), reactions can be included/excluded and their reversibility be modified. Similarly, metabolites can be included/excluded from balancing in the "Metabolites" frame (center). The "Flux Ratios" frame (right) contains the relevant flux ratios as computed by the **ratio** module, which can also be included/excluded via the user interface. Finally, the experimentally determined reaction rates and other metabolic parameters have to be specified in the "Extracellular fluxes" text field at the bottom of the **netto** window.

Then, having all required data available, flux distributions can be computed. The flux ratios calculated within the **ratio** and the mass balances of the metabolites of the reaction network build a linear equation system, which is made mathematically accessible by transformation to matrix notation. If the equation system is determined or even overly constrained, it is simply solved by a least square optimization. In case the system is underdetermined, `netto` can either estimate all calculable fluxes or optimize the solution with regard to a particular flux or intermediate. In any case, the user has to start the computation by pressing one of the buttons in the "Analysis" frame of the `netto` window. When the calculation has finished, an F value that indicates the quality of the obtained solution is displayed within the frame. If the user considers the F value being too large, he can either modify the network model and restart the calculation, or simply re-run the calculation with the same network configuration, since the computations are partly based on random numbers and can thus lead to different results also without parameter changes.

Summarizing, FiatFlux is a highly interactive desktop software that provides a graphical user interface to a Matlab implementation of METAFoR and ^{13}C-based metabolic flux analysis. For an in-depth explanation of the software and the calculations the reader is referred to [95] and [230].

5.2 FiatFlux-P Workflows

Naturally, interactive workflows become limiting when hundreds or even thousands of data sets have to be handled. The previous chapter showed how the highly interactive GeneFisher web application has been turned into a flexible workflow realization, called GeneFisher-P. Metabolic flux analysis with Fiat-Flux is a similarly interactive process that has not been designed for batch operation. Like GeneFisher-P, *FiatFlux-P* [90] is a workflow-based version of the original interactive analysis software that allows the user to work with the FiatFlux functionality in a highly flexible and automated manner. It has been developed in the scope of a joint effort with the Laboratory of Chemical Biotechnology at the TU Dortmund, where it has actively been used for large-scale data analyses (cf. [89]).

Among the Bio-jETI applications, FiatFlux-P is the application where the speedup of the analyses was most considerable: the analysis with FiatFlux-P runs about three to five times faster than the corresponding manual analysis. That is, a complete ^{13}C-based metabolic flux analysis performed with FiatFlux-P requires about four minutes instead of 12 to 20 minutes needed for the manual analysis of a single data set. As metabolic flux experiments do not only produce a single data set that has to be analyzed, but often 20, 50 or even 150 data sets, this means that the time spent for the data analyses for an experiment is now only about 1:20 h, 3:20 h, or 10 h instead of up to 6:40 h, 16:40 h, or 50 h, respectively. Furthermore the manual analysis requires the full attention of an (experienced) human user, hence it is expensive in the sense that it can easily consume a whole man-week of work. In contrast, the automatic analysis workflows run autonomously in the background, possibly also overnight, so that the researcher can deliberately focus on other tasks in the meantime.

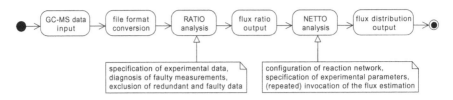

Fig. 5.5 Abstract FiatFlux workflow

As detailed in Section 5.1.3, FiatFlux consists of the modules `ratio` and `netto`, where the former performs the flux ratio computation and the latter calculates the net flux distributions. Figure 5.5 on the next page summarizes the general FiatFlux workflow: The input data (GC-MS data in netCDF format, the output format of many MS devices) is first converted into the internal FiatFlux (.ff) data format, then the `ratio` and `netto` analyses are performed and their respective results are output. In the original FiatFlux software, both `ratio` and `netto` require several user interactions via the graphical user interface (GUI) FiatFlux. These are indicated by the two note boxes at the bottom of the figure.

Preparing GeneFisher for its realization as GeneFisher-P was comparatively simple: its building blocks had to be integrated as services in Bio-jETI and combined to (new) workflows. Essentially, only suitable "glue code" was needed. With FiatFlux, however, the case is not as easy. The user interactions of FiatFlux require specific expert knowledge, as GC-MS data quality and relevance have to be assessed and the resulting data has to be compared with biochemical knowledge in the framework of metabolic flux analysis. Consequently, this expert knowledge had to be translated into quantifiable criteria, which could be used for the automated determination of intracellular flux distributions. Thus, for turning FiatFlux into FiatFlux-P it was not sufficient to turn its components into services, but it was also necessary to implement functionality that emulates the interaction with an expert user as close as possible.

For analysis procedures that do not involve human interaction, it is easy to see that the automation of the in silico experiment using workflow technology increases the speed of the analyses without influencing the results at all. However, also workflow realizations of usually interactive analysis processes do not necessarily impact the quality of the results: it is often possible to identify quantifiable criteria in the human expert's analysis behavior, and apply these for at least heuristic user interaction emulation. In the case of FiatFlux-P, most automatically obtained results were as good as the manually acquired ones (cf. [89, Chapter 6], [90]). This means in particular that in such cases automated experiments can be used for a (fast) pre-screening or initial analysis of large amounts of data, and only the remaining "difficult" data sets have to undergo (time-consuming) manual analysis.

5.2.1 FiatFlux Services

Turning FiatFlux' components into Bio-jETI-compliant services required some major changes to its Matlab code base, in particular the rigorous removal of all code for graphical components and the replacement of invocations from the GUI by (parameterized) functions that can be called from an external application. The user interactions required during the FiatFlux analysis process can be categorized into simple input of experimental data or the selection of modeling parameters and more intricate user interactions targeted to the optimization of the calculations. Whereas the data input could easily be replaced by parameterized functions, for the other steps the expert logic had to be translated into quantifiable criteria. The functions developed to emulate these tasks are packaged into two new modules `ratio_guiemulation` and `netto_guiemulation`, respectively (detailed descriptions of the emulation functions and some other additionally implemented functionality can be found in [89, Chapter 6].

Next, the newly implemented functions were combined with the standard FiatFlux functions in order to provide self-contained functional units. These were then integrated and turned into SIBs using the jETI technology (cf. Section 2.1.2). The following SIBs are central in FiatFlux-P:

- `ConvertMS` converts mass spectrometry data from netCDF format to the FiatFlux format.
- `Ratio` performs a predefined, complete flux ratio computation.
- `Netto` performs a predefined, complete net flux distribution computation.
- `Netto_CustomModel` performs a predefined, complete net flux distribution computation based on a user-defined network model.
- `Netto_JointRatios` performs a predefined, complete net flux distribution computation based on the combined flux ratios from two different experiments.

At the beginning of the project, a larger number of services was used, which provided more fine-granular analysis steps. In the course of the project, however, it turned out that a quite coarse-granular service library, which provides predefined variants of the major analysis steps, rather than exposing computational details of the basic analysis steps to the workflow level, is more adequate.

5.2.2 FiatFlux Workflow for Single Data Sets

With these SIBs and the standard set of SIBs provided by Bio-jETI it was now possible to define appropriate FiatFlux workflows. Figure 5.6 shows a workflow for the analysis of a single data set, analogous to the abstract workflow depicted in Figure 5.5: The input CDF file is read and converted into the FiatFlux format, and then used for flux ratio estimation and calculation of net fluxes. In between the respective results are saved into text files for later use.

Fig. 5.6 FiatFlux worfklow for the analysis of a single data set

5.2.3   Batch Processing FiatFlux Workflows

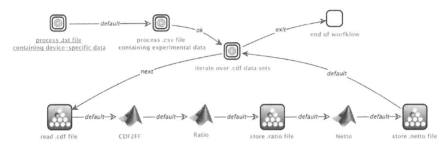

Fig. 5.7 METAFoR and ^{13}C-based metabolic flux analysis of several data sets

The actual objective of the FiatFlux-P project was to automate the Fiat-Flux analysis procedure in order to obtain a standardized analysis process and to increase the amount of data sets that can be analyzed in a certain amount of time, that is, to implement a workflow that allows for batch processing of numerous .cdf data sets. In FiatFlux, all experimental data have to be entered manually by the user at different steps of the analysis procedure and at different parts in the GUI. Figure 5.7 shows a workflow for the processing of several data sets: The user has to specify the working directory and a .csv file that lists the .cdf files under consideration and all experimental parameters. (The .csv format can be exported from all common spreadsheet programs, thus researchers can continue to document their experiments within MS Excel, OpenOffice Calc or other.) The batch processing workflow then simply repeats the single data set analysis workflow described above, processing another data set in each iteration, until all input files have been analyzed. As user input is only required once at the beginning, this workflow is able to process very large sets of input data autonomously, speeding up the analysis procedure significantly.

5.2.4   Further Variations of the FiatFlux-P Workflow

Instead of the standard Netto service that has been used in the examples above, the Netto_CustomModel and Netto_JointRatios variants can be included in the analysis workflows. For the former, it is additionally required to provide a custom network model file, whereas for the latter a second ratio result has to be provided. Both these variations have been extensively used in the Laboratory of Chemical Biotechnology, TU Dortmund, for the analysis of various experimental data.

As graphical presentation of metabolic flux distributions greatly facilitates their interpretation, the FiatFlux code has furthermore been extended by functionality for exporting the calculated fluxes in a format that can be used for visualizing the calculated metabolic flux distributions using the OMIX software [88], an editor for drawing metabolic reaction networks. More precisely, the reaction rates together with specific reaction identifiers are exported to a .csv file, which can be saved from within the workflow just as the `ratio` and `netto` result files.

This format can later on easily be interpreted by an OVL (OMIX Visualization Language) script in order to equip default network diagrams with markups according to the obtained results. The customized metabolic flux charts can then be exported by OMIX into different bitmap and vector graphic formats such as .png, .jpg and .svg. Within the FiatFlux-P project, ready-to-use OMIX network diagrams for a number of frequently analyzed model organisms are provided, along with an OVL script offering two different markup variants, one for the visualization of a single result data set, where the line width of the reaction arrows is adjusted to the specific flux, and another for the visualization of multiple result data sets, where the values of the reaction rates are assigned to the arrows representing the respective reaction.

5.3   Constraint-Driven Design of FiatFlux-P Workflows

In the previous examples, the tools and algorithms that were involved were mostly more or less common, third-party services, and thus it was indicated to automatically retrieve the basic domain information from third-party knowledge repositories, too. FiatFlux-P, in contrast, is a very specific application, tailored to the development of workflows for metabolic flux analysis with a particular variant of the FiatFlux software. Hence the EDAM ontology does not comprise the suitable terminology for this application. However, FiatFlux-P uses a comparatively small set of services and comprises only a small number of different data types, so that setting up a clear and unambiguous domain model is in fact straightforward. In particular, the service and type taxonomies have been defined manually and closely tailored to the specific application examples, providing exactly the level of generalization and refinement that is required for this particular application. The corresponding domain model for FiatFlux-P is presented in the following, before an exemplary workflow composition problem is discussed.

5.3.1 Domain Model

As for any PROPHETS workflow scenarios, also the domain model for FiatFlux-P comprises service and type taxonomies, service descriptions and domain constraints. They are described in the following.

Taxonomies

Figure 5.8 shows the service taxonomy of the FiatFlux-P domain model. The taxonomy defines the service categories *DataInput*, *DataOutput*, *FiatFlux* and *Helpers*, of which *FiatFlux* defines the additional subclasses *Preprocessing*, *MetabolicFluxAnalysis* and *MetaforAnalysis* in order to distinguish between the different analysis steps. As the figure shows, the concrete services (available as SIBs) are then sorted into the taxonomy tree, that is, the different file reading services are classified as data input services, the file writing service is classified as data output service, the different FiatFlux services are linked to the corresponding sub-categories, and the special service for iterating over multiple CDF data sets is classified as a helper service.

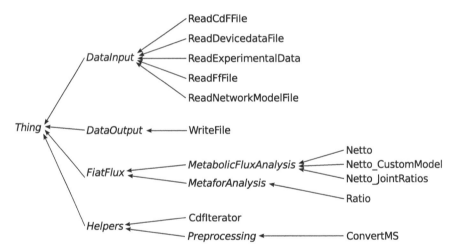

Fig. 5.8 Service taxonomy of the FiatFlux-P domain model

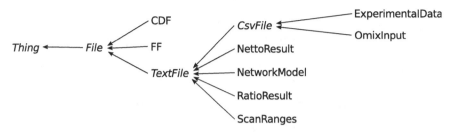

Fig. 5.9 Type taxonomy of the FiatFlux-P domain model

The type taxonomy of the FiatFlux-P domain model is shown in Figure 5.9. The FiatFlux-P workflows read their input data from files and also write all (intermediate and final) results into files, hence all types in the domain model are subclasses of *File*. The CDF files (containing the raw data from the GC/MS measuring device) and the FF files (the internal FiatFlux format) are not described further. The *TextFile* subclass comprises the involved file types that contain textual content, such as the analysis results and additional input data. Among them, *CsvFile* is a special kind of text file that contains textually represented values in comma-separated format, which is a simple way to store tables. Within FiatFlux-P, .csv files are used for the documentation of the experimental parameters belonging to a CDF data set, and for exporting analysis results in a format that can be interpreted by the OMIX visualization software (cf. Section 5.2).

Services

Table 5.1 lists the services contained in the FiatFlux-P domain model (in alphabetical order) and displays their input and output type annotations. Besides the FiatFlux services that process the different kinds of experimental data and produce different results, the domain model comprises services that read certain kinds of files (not requiring any inputs), the file-writing service that stores all kinds of files (but does not itself produce a new file), and the CdfIterator that processes a collection of experimental data sets.

Constraints

Using the constraint templates of the PROPHETS constraint editor, domain constraints have been defined that cover general properties of FiatFlux-P workflows, such as that the Netto analyses have always be preceded by a Ratio analysis, that particular services should not be used more than once in order to prevent unintended repeats, that data that is read is used subsequently, or that analysis results should finally be stored:

- Enforce the use of *FiatFlux*.

Table 5.1 Services in the FiatFlux-P domain model

Service	Input types	Output types
CdfIterator	ExperimentalData	
ConvertMS	CDF	FF
Netto	FF, ScanRanges	FF, NettoResult, OmixInput
Netto_CustomModel	FF, ScanRanges, NetworkModel	FF, NettoResult, OmixInput
Netto_JointRatios	FF, ScanRanges	FF, NettoResult, OmixInput
Ratio	FF, ScanRanges	FF, RatioResult
ReadCdfFile		CDF
ReadExperimentalData		ExperimentalData
ReadNetworkModelFile		NetworkModelFile
ReadScanRangesFile		ScanRanges
WriteFile	*File*	

- Do not use `ReadExperimentalDataFile`.
  Do not use `CdfIterator`.
- *MetabolicFluxAnalysis* depends on *MetaforAnalysis*.
- Do not use *MetabolicFluxAnalysis* more than once.
  Do not use `ReadNetworkModelFile` more than once.
  Do not use `ReadScanRangesFile` more than once.
- If `ReadCdfFile` is used, `ConvertMS` has to be used next.
- If `ReadNetworkModelFile` is used, `Netto_CustomModel` has to be used subsequently.
- If *FiatFlux* is used, *DataInput* must not to be used subsequently.
  If *DataOutput* is used, *FiatFlux* must not to be used subsequently.
- Use *WriteFile* as last service in the solution.

Furthermore, it is not desirable to use `ConvertMS` or `Ratio` arbitrarily often. However, is also not desirable to prevent the their repeated use of in general, since in case `Netto_JointRatios` is used, for instance, two input data sets have to read processed by these services. Note that as it is not possible to express constraints like "service X has to be used exactly twice" or "service Y must not be used more than twice" with the available general constraint templates, they are not included in the exemplary domain model from above. An experienced user could, however, add the following formulae to the domain model using the "expert mode formula" template:

- If `Netto` or `Netto_CustomModel` is used for the *MetabolicFluxAnalysis* step, use `ConvertMS` and `Ratio` exactly once.
  $F(\langle \mathsf{Netto} \rangle true \vee \langle \mathsf{Netto\_CustomModel} \rangle true) \Rightarrow$
  $(F(\langle \mathsf{ConvertMS} \rangle true) \wedge G(\langle \mathsf{ConvertMS} \rangle true \Rightarrow X(G(\neg \langle \mathsf{ConvertMS} \rangle true))) \wedge$

  $F(\langle \mathsf{Ratio} \rangle true) \wedge G(\langle \mathsf{Ratio} \rangle true \Rightarrow X(G(\neg \langle \mathsf{Ratio} \rangle true))))$

- If `Netto_JointRatios` is used, use `ConvertMS` and `Ratio` exactly twice.

  $F(\langle\text{Netto\_JointRatios}\rangle true) \Rightarrow$
  $(F(\langle\text{ConvertMS}\rangle true \wedge X(F(\langle\text{ConvertMS}\rangle true)) \wedge$
  $G(\langle\text{ConvertMS}\rangle true \Rightarrow X(G(\langle\text{ConvertMS}\rangle true \Rightarrow X(G(\neg\langle\text{ConvertMS}\rangle true))))) \wedge$
  $F(\langle\text{Ratio}\rangle true \wedge X(F(\langle\text{Ratio}\rangle true)) \wedge$
  $G(\langle\text{Ratio}\rangle true \Rightarrow X(G(\langle\text{Ratio}\rangle true \Rightarrow X(G(\neg\langle\text{Ratio}\rangle true)))))$

As in the previous examples, problem-specific constraints can be defined in addition to the constraints of the domain model during the workflow design phase to narrow the possible solutions further, for instance by requiring that particular services should not be used at all or that particular results should finally be available.

5.3.2 Exemplary Workflow Composition Problem

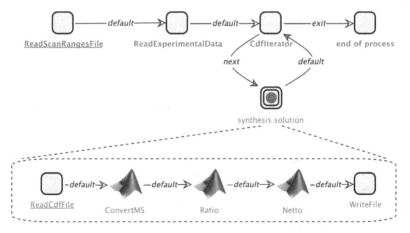

Fig. 5.10 Loose programming of a batch-processing FiatFlux-P workflow

Based on the domain model as described above, it is now possible to design FiatFlux-P workflow (semi-) automatically. However, due to the loop that is required for repeating the original FiatFlux-P workflow in the batch-processing variant, this workflow can not be created completely by the current synthesis algorithm, which is restricted to producing linear sequences of services. It is, however, possible to predefine a sparse workflow model, in which the looping behavior and other crucial parts are manually predefined, and to subsequently fill in linear parts of the workflow automatically. Hence, the synthesis problem is to create the loosely specified body of the FiatFlux-P batch processing loop as shown Figure 5.10 (top).

The scan ranges and experimental data are read before the loop that iterates over the CDF data sets is entered (cf. the original batch-processing FiatFlux-P workflow in Section 5.2). This is recognized by PROPHETS' data-flow analysis mechanism, so that the synthesis is aware that services generating them do not have to be included in the solution. The reading of the CDF input data and the storing of the results is however not yet contained in the workflow. While the synthesis will recognize the necessity of including the reading of input data from the interface specifications, the inclusion of WriteFile is ensured by a domain constraint. Figure 5.10 (bottom) shows one of the shortest solutions to this synthesis problem, where the CDF file is read, the FiatFlux analysis steps performed, and the result finally stored in a file.

By varying the constraints it is of course possible to obtain other solutions to these synthesis problems, for example corresponding to the variations described in Section 5.2. For instance, a constraint that enforces the use of Netto_CustomModel would not only make the synthesis include the respective SIB, but also the ReadNetworkModel SIB in order to have all required input data available.

Again, the FiatFlux-P domain is comparatively small example, where furthermore many general constraints are already contained in the domain model. Hence the number of solutions is usually easily manageable and the user only has to provide minor additional hints concerning his intentions about the workflow in order to obtain the desired solutions.

With no constraints applied and searching for solutions up to length 7, for instance, more than 5,000 solutions are found, while the domain constraints reduce the number of returned solutions to 44. Additionally enforcing the use of Netto_CustomModel leaves mere 4 solutions. Note that the (one) minimal solution for this synthesis problem is found in depth 6 and consists of ReadCdfFile, ConvertMS, ReadNetworkModelFile, Ratio, Netto_CustomModel and WriteFile. The additional three solutions that are found in depth 7 contain ConvertMS or Ratio twice or include the actually not required ReadScanRanges.

6

Microarray Data Analysis Pipelines

This fourth and final example considered in this book deals with workflows for the analysis of microarray data based on the statistical methods provided by Bioconductor [105]. This kind of bioinformatics workflows is frequently referred to as data analysis "pipelines", as they typically have a simple, in fact mostly linear, workflow structure, while the (often considerable) complexity of the individual analysis steps is encapsulated by the services. Consequently, the constraint-driven workflow composition functionality of the PROPHETS plugin, which is based on a linear-time logic synthesis algorithm, can conveniently be used for complete, start-to-end generation of the workflows.

6.1   Background: Microarray Data Analysis

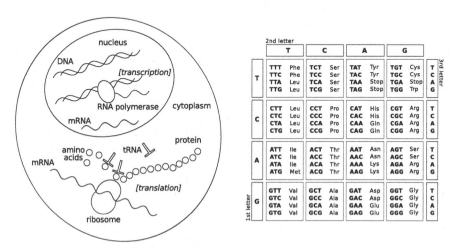

Fig. 6.1 Protein biosynthesis and the genetic code

In order to make a living organism, the genetic information (i.e., the DNA) has to be interpreted and processed into organic substances. Figure 6.1 (left) gives a schematic picture of this process, which is also called *protein biosynthesis*. The first step in this process is called *transcription*: an enzyme (the RNA polymerase) produces an RNA copy of a part of DNA. This copy, the *messenger RNA* (mRNA), leaves the nucleus and goes into the cytoplasm of the cell, where it is in the subsequent *translation* step used as template for the synthesis of proteins at the ribosomes. The ribosomes "read" the sequences in triples (called *codons*) and translates them into single amino acids according to the *genetic code* (right side of the figure). The matching amino acids are carried to the ribosomes by the *transfer RNA* (tRNA) molecules and added to the protein under construction one after another.

Although all cells of an organism have the same genetic material (the same genotype), they appear in different forms (phenotypes), differing in physiology and morphology. The reason is that not all genes, but only those that are currently needed, are transcribed (expressed) and used as RNA or further translated into proteins, and hence the set of expressed genes (the transcriptome) varies.

6.1.1   Systems Biology: Transcriptomics

Accordingly, *transcriptomics*, another sub-discipline of Systems Biology, is concerned with "global analyses of gene expression with the help of high-throughput techniques such as DNA microarrays" [279, Glossary]. The following introduction to its principles is largely based on [279, Section 7.1.1].

Transcriptomics aims at studying and understanding the regulation and expression of genes, which can provide insight into the functions of the gene products. Typical gene profiling studies are concerned with the comparison of the gene expression patterns of different cell populations. For instance, the gene expression profiles of healthy and tumor cells can help to characterize and classify the tumor properly, which is necessary for finding the optimal treatment for the patient.

DNA microarray technology allows for high-throughput, time-efficient analysis of whole transcriptomes of cells. Figure 6.2 gives a schematic overview of a microarray experiment: DNA microarrays consist of some sort of solid support material (e.g., a glass slide or a nylon membrane), on which thousands of nucleic acid spots are arranged close to each other. Each spot contains many copies of a unique, single-stranded DNA fragment, which can be unambiguously assigned to a specific gene. During hybridization, the RNA that has been isolated from the cell population under investigation is applied to the DNA microarray, and the RNA fragments pair with the spots that carry the complementary single-stranded DNA molecule. After washing (for cleaning the microarray from of the unbound, left-over RNA), the DNA microarray is put into a microarray scanner in order to detect the intensity of the spots, which corresponds to the amount of RNA that has hybridized.

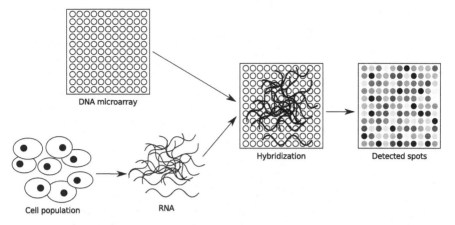

Fig. 6.2 Microarray experiment

There are basically two types of DNA microarrays: high-density oligonu-cleotide (HDO) arrays and complementary DNA (cDNA) microarrays. HDO microarrays provide high-quality spots of short sequences of 20-50 oligonu-cleotides length, which are arranged on the carrier material in extreme density. These arrays are comparatively expensive, but allow for the identification of absolute quantities of hybridized RNA. Note that market leader Affymetrix calls its HDO microarrays *GeneChips*, and this term is also frequently used in literature. The spots of cDNA microarrays consist of longer cDNA probes, which are hybridized with the differently labeled (red and green dyes) cDNA probes of two cell populations. The resulting two-color spotted cDNA mi-croarrays only allow for the identification of relative quantities, but they are more economical than HDO microarray experiments. Similarly, protein ar-rays are used in proteomics (another sub-discipline of Systems biology) for the simultaneous functional analysis of proteins.

A proper experimental design is crucial for microarray experiments in order to minimize systematic errors, such as measurement deviations due to incorrect instrument calibrations or changing environment conditions. However, also for well-performed experiments, the analysis of the results (mostly based on the detected spot intensities) is again complex, as detailed in the next section.

6.1.2 Microarray Data Analysis

The very first step in the interpretation of data from microarray experiments is the analysis of the produced displays, that is, the measuring of the spots' intensities and their conversion into numerical values. This demanding image processing task (spots have to be identified unambiguously) is usually carried out by the specialized software of a microarray scanner, so that the researcher starts the actual, specific data analysis based on the spot intensities and possibly available meta-data. The preprocessing and differential expression analyses that are usually carried out next are mainly statistical methods. This section gives a brief introduction to the concepts of preprocessing and statistical analysis of microarray data. For more detailed information, the reader is referred to, e.g., [48, 125].

Due to the measurement variations caused by systematic and stochastic effects, raw data from microarray experiments needs to be preprocessed before statistical analyses of the data can be performed. As described in [48, Section 1.4], the intensity Y of a single probe on a microarray is generically modeled by:

$$Y = B + \alpha S$$

where B stands for the background noise (optical effects, non-specific binding), α is a gain factor, and S the amount of measured specific binding (including measurement error and probe effects). According to the additive-multiplicative error model for microarray data, the measurement error is usually modeled by

$$log(S) = \theta + \phi + \epsilon$$

where θ is the logarithm of the true abundance of the molecule, ϕ is a probe-specific effect and ϵ the measurement error. These models, or similar ones, are the basis for various preprocessing methods.

Typically, the preprocessing of microarray data involves three principal steps (cf. [48, Section 1.2]):

1. *Background adjustment* aims at estimating the effects of non-specific binding and noise in the optical detection on the measured probe intensities in order to adjust the measurements of specific hybridization accordingly.
2. *Normalization* adjusts the results further to make experiments of different array hybridizations comparable.
3. *Summarization* combines the background-adjusted and normalized intensities of multiple probes into one quantity that estimates the amount of RNA transcript.

There are various methods for all three steps available (cf., e.g., [48, Chapters 1–6] for an overview), and naturally, there are also preprocessing strategies that involve additional steps (cf., e.g., [48, Chapter 2]). After preprocessing, a matrix of expression values is available as basis for subsequent analyses.

A common first processing step carried out on the expression matrix is the filtering of the expression values according to some (quality) criterion, in order to take only samples with specific properties into account. Frequently applied subsequent statistical analysis steps are, for instance, differential expression analysis (aiming at the identification of genes for which expression is significantly different between the samples, cf. [48, Chapter 14]), and Cluster analysis (aiming at the recognition of patterns in gene expression profiles, cf. [48, Chapter 13]).

Furthermore, (automated) annotation of data and analysis results can be applied to complement the statistical analyses in order to enrich the experimental data with knowledge from external repositories (cf., e.g., [48, Chapter 7]). For instance, literature databases like PubMed [322] can be searched for finding articles related to the identified differentially expressed genes, the corresponding parts of the Gene Ontology [32] can be taken into account, or pathway databases like KEGG [146] or cMAP [56] can be used for associating experimental findings with the related pathways.

6.1.3 *Bioconductor*

As detailed in the previous section, the interpretation of microarray data is predominantly carried out via statistical analysis methods. Different commercial and academic software frameworks provide support for performing statistical analyses of microarray data. In academia, the statistics language GNU R [18] in conjunction with the specialized R packages provided by the Bioconductor project [105, 1] is particularly popular.

GNU R is a widely used programming language and software environment for statistical data analysis and visualization. R has a rich collection of standard functions for commonly required functionality, covering, for instance, linear and nonlinear modeling, classical statistical tests, classification and clustering. R's range of functionality can easily be extended by user-created packages. Packages can be made available to the user community by submitting them to one of the R package repositories, such as the Comprehensive R Archive Network (CRAN) [16] for all kinds of packages, or Bioconductor specifically for bioinformatics packages. Bioconductor comprises comprehensive libraries of functions and meta-data predominantly for the analysis of data from high-throughput genomics and molecular biology experiments, and additionally provides several example data sets that are useful for testing, benchmarking and demonstration purposes.

Various (parts of) microarray data analysis procedures have been described in literature (cf., e.g., [48, 125]) and in the reference manuals and additional manuscripts provided with the packages at the Bioconductor web site [1]. In order to illustrate how these analyses can be carried out using R and Bioconductor, this section discusses a simple example microarray data analysis procedure (inspired by [48, Chapter 25]) that uses several common Bioconductor packages. Starting point for the analysis is a set of Affymetrix CEL files [22] that have been obtained from a microarray scanner (cf. Section 6.1.1). After preprocessing and filtering of these raw probe-level data, a differential expression analysis is carried out. Finally, the literature database PubMed [322] is queried for relevant articles using the names of the top differentially expressed genes as search keywords.

```
1   # load required packages
2   library (" Biobase ")
3   library (" affy ")
4   library (" AffyExpress ")
5   library (" limma ")
6   library (" hgu95av2 . db ")
7   library ("XML")
8   library (" annotate ")
9
10  # load input data
11  celpath <- " spikein /"
12  phenodatafile <- " spikein_pdata . txt "
13  phenodata <- read . table ( phenodatafile ,  row . names = 1,  header
        = TRUE,  sep = "")
14  affybatch <- ReadAffy ( celfile . path=celpath ,  phenoData =
        phenodata )
15
16  # AffyExpress : preprocessing
17  exprset <- pre . process ( method = " rma ",  raw = affybatch ,  plot
        = FALSE)
18
19  # AffyExpress : filtering
20  filteredexprset <- Filter ( exprset ,  numChip = 3,  bg = 7.0)
21
22  # differential expression analysis using limma
23  fit <- lmFit ( exprset )
24  fit <- eBayes ( fit )
25  toptable <- topTable ( fit ,  adjust = " fdr ")
26
27  # get PubMed abstracts for top genes
28  genenames <- as . character ( toptable$ID )
29  abstracts <- pm . getabst ( genenames ,  " hgu95av2 ")
30  abstracts <- unlist ( abstracts ,  recursive = FALSE)
31  pmAbst2HTML ( abstracts ,  filename = " spikein −abstracts . html ",
        frames = FALSE)
```

Fig. 6.3 R script performing a simple microarray data analysis

The code listed in Figure 6.3 shows an R script that performs the described analysis steps. At first, the required packages are loaded: **Biobase** is the fundamental package of Bioconductor [105], containing standardized data structures for the representation of genomic data. Most importantly here, it defines the **ExpressionSet** class [93], which conveniently manages different data from microarray experiments: **assayData** (the actual expression data), **phenoData** (meta-data on the samples), **featureData** and **annotations** (meta-data on the used chip or technology), **protocolData** (technical information about sample processing) and **experimentData** (flexible description of the experiment). The **affy** package [102] provides functionality for the processing of Affymetrix GeneChip data, for instance a function for reading the raw data from Affymetrix GeneChip experiments into an **AffyBatch** object that serves as basis for the subsequent preprocessing steps. **AffyExpress** [350] contains

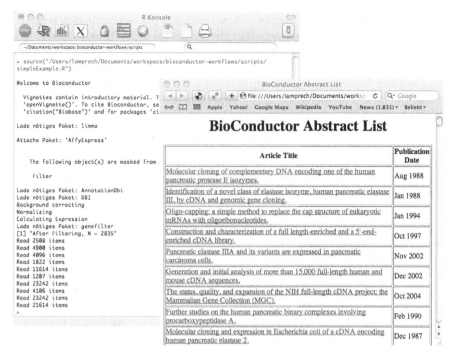

Fig. 6.4 Execution of the example script and display of results

a number of easy-to-use wrapper functions for often recurring analysis steps when dealing with Affymetrix array, in particular regarding preprocessing and expression value filtering. The `limma` package [290] provides functionality for the analysis of gene expression data using linear models. The last three packages that are loaded are required for the annotation step, that is, the search for relevant PubMed articles: the `hgu95av2.db` annotation data package [63] contains annotation data for the Affymetrix chip hgu95av2 (Human Genome U95), the `annotate` package [103] provides functions to associate meta-data of different kind and from different sources with the analyzed experimental data, and `XML` [181] is simply required for proper HTML formatting.

When the required R and Bioconductor packages have been loaded, lines 11–14 load the input data. The CEL files are located in the `spikein/` directory and `spikein_pdata.txt` is the text file containing the corresponding pheno data (i.e. meta-data on the CEL data sets). The pheno data are read using the standard R function for reading table data, while the function `ReadAffy` from the `affy` package is used to create an `AffyBatch` object from the pheno data and the CEL data. Next (line 17), the `pre.process` function of the `AffyExpress` package is called for preprocessing the data in the `AffyBatch` object and creating the corresponding `ExprSet`. Using the `Filter` function of the `AffyExpress` package, line 20 filters the contained data sets so that only those remain in the `ExprSet` that have at least 3 chips with a background value of 7.0 or higher. Then (lines 23–25), a simple differential expression analysis is carried out using the `limma` package. More precisely, it calls the `lmFit` and `eBayes` (fitting a linear model to the expression data and ranking the genes in order of evidence for differential expression, respectively) and creates a `TopTable` object containing the top-ranked genes. Finally (line 28 ff.), the names of the genes in the `TopTable` are extracted and used for a PubMed query. The results are written into an HTML file in table format.

The script can directly be executed, for instance via the R console as shown in Figure 6.4 (left). After successful execution, the HTML table containing the links to the related PubMed articles is available at the local file system and can be opened in any standard browser, as shown at the right side of the figure. Note that as R is an interpreted language, the commands contained in the script can also be directly entered into the R console one after another.

Interestingly, working with Bioconductor essentially consists of accessing predefined functions from the plethora of available packages. Thus, developing R scripts based on Bioconductor packages is indeed rather service-level orchestration of existing software building blocks than actual programming of (new) functionality. Also higher-level objects like the `AffyBatch` or `ExpressionSet` classes provide convenient, user-level data objects by abstracting from the internal details of data handling. Being conceived as programming libraries in the first place, however, most Bioconductor packages provide relatively fine-grained functionality, which keeps the development of Bioconductor programs at a lower, programming-language level and demands a good deal of R knowledge from the user.

6.2 Microarray Data Analysis Workflows

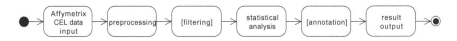

Fig. 6.5 Abstract microarray data analysis pipeline

This last scenario demonstrates how biostatistics workflows for the analysis of microarray data can be designed with Bio-jETI, making use of different Bioconductor libraries (cf. Section 6.1.3) in the underlying services. As sketched in Section 6.1.2 and shown in Figure 6.5, microarray data analysis basically consists of a sequence of statistical analysis steps, possibly supplemented by data annotation based on external knowledge. However, the actually applied analysis and annotation steps vary considerably, as they depend on both the nature of the input data as on the analysis objectives. Accordingly, the abstract microarray analysis pipeline depicted in Figure 6.5 gives only a rather simplified characterization of the workflow variants: While preprocessing and statistical analysis are usually applied to each input data set, filtering and annotation steps are useful in most cases, but not mandatory in general (indicated by the square brackets around the step descriptions in the figure). Note that whereas usually one data loading and preprocessing step per workflow is sufficient, there may well be applications where more than one filtering, statistical analysis or annotation step is applied, and where then also different results are obtained at different stages of the workflow.

6.2.1 Bioconductor Services

Having identified principal units of functionality for the analysis of microarray data using Bioconductor packages, the following services (among others) were provided for this application using the jETI technology (cf. Section 2.1.2):

- `AffyExpress_filter1` applies a filter function to an ExpressionSet so that at least a certain number of chips are greater than a given background [350].
- `AffyExpress_filter2` applies a filter function to an ExpressionSet so that the range of a gene has to be greater than a given value (range) [350].
- `AffyExpress_preprocess` performs preprocessing of Affymetrix microarray data using the `pre.process` function of the AffyExpress package [350].
- `Annaffy_aafTableAnn` creates a table given a set of probe IDs and desired annotation types [286].
- `Annaffy_aafTableInt` creates a table containing expression values given an ExpressionSet. In the resulting HTML table, the expression values will have backgrounds with varying intensities of green depending on the expression measure [286].
- `CreateExpressionSet` creates an ExpressionSet object from expression values and phenodata.
- `DifferentialExpressionAnalysis_ReplicateArrays` performs a basic differential expression analysis, suitable for a set of replicate arrays using the limma package [290].
- `Expresso` performs preprocessing of Affymetrix GeneChip data [102]. Background correction, normalization, PM correction and summarization methods are configurable.
- `GCRMA` performs preprocessing Affymetrix GeneChip data [351].
- `Genefilter_CV` applies a filter function for the coefficient of variation to an ExpressionSet [104].
- `Genefilter_kOverA` applies a filter function for k elements larger than A to an ExpressionSet [104].
- `Genefilter_maxA` applies filter function to filter according to the maximum to an ExpressionSet [104].

- `Genefilter_pOverA` applies a filter function to filter according to the proportion of elements larger than A to an ExpressionSet [104].
- `GetPubMedAbstracts` gets the PubMed abstracts for the genes in a TopTable [103].
- `RMA` performs preprocessing of Affymetrix GeneChip [102].
- `Threestep` performs preprocessing of Affymetrix GeneChip data [102, 47]. Background correction, normalization and summarization methods are configurable.

In addition, some SIBs are provided that can be used to load example microarray data sets directly from the server, which is convenient when just testing or experimenting with the workflows, as no (likely lengthy) data upload is required:

- `LoadDilutionBenchmarkData` loads a benchmark data set of a dilution microarray experiment. The data set is described in [73] as follows: "Two sources of cRNA, human liver tissue and central nervous system (CNS) cell line, were hybridized to human arrays (HG-U95Av2) in a range of dilutions and proportions [136]. We studied data from six groups of arrays that had hybridized liver and CNS cRNA at concentrations of 1.25, 2.5, 5.0, 7.5, 10.0 and 20.0 μg total cRNA. Five replicate arrays were available for each generated cRNA (n = 60 total). Oligos from genes specific to foreign species were synthesized and added to each hybridization mixture at nominal amounts of 0.5, 1, 1.5, 2, 3, 5, 12.5, 25, 50, 75 and 100 pM. These oligos correspond to probe-sets: BioB-5, BioB-M, BioB-3, BioC-5, BioC-3, BioDn-5 (all Escherichia coli), CreX-5, CreX-3 (phage P1), and DapX-5, DapX-M, DapX-3 (a Baccillus subtilis gene). Oligos corresponding to the 3'-end, middle and 5'-end of each gene were synthesized and added separately. The same concentrations were used across all 60 arrays."
- `LoadSpikeInBenchmarkData` loads a benchmark data set of a spike-in microarray experiment. The data set is described in [73] as follows: "Human cRNA fragments matching 16 probe-sets on the HGU95A GeneChip were added to the hybridization mixture of the arrays at concentrations ranging from 0 to 1024 pM. The same hybridization mixture, obtained from a common tissue source, was used for all arrays. The cRNAs were spiked-in at a different concentration on each array (apart from replicates) arranged in a cyclic Latin square design with each concentration appearing once in each row and column. The design is described in detail by [135]."
- `LoadHgu133SpikeInBenchmarkData` loads a benchmark data set of a spike-in microarray experiment. The experimental procedure was the same as for `LoadSpikeInBenchmarkData`, except that the experiment was carried out on the HG-U133Atag chip instead of the HGU95A GeneChip.

These benchmark data sets are used by the `affycomp` package [137] for evaluating and comparing microarray preprocessing methods. Selected assessment functions are also available as SIBs:

- `DilutionAssessmentBasic` assesses a single preprocessing method based on its application to the dilution benchmark data set.
- `DilutionAssessmentComparative` assesses and compares two preprocessing methods based on their application to the dilution benchmark data set.
- `SpikeInAssessmentBasic` assesses a single preprocessing method based on its application to a spike-in benchmark data set.
- `SpikeInAssessmentComparative` assesses and compares two preprocessing methods based on their application to a spike-in benchmark data set.
- `SpikeInAssessment2Basic` assesses a single preprocessing method based on its application to a spike-in benchmark data set using an alternative assessment function.
- `SpikeInAssessment2Comparative` assesses and compares two preprocessing methods based on their application to a spike-in benchmark data set using an alternative assessment function.

Note that the SIBs in this library call R scripts that bundle particular Bioconductor functionality, following the approach that has already been taken for the case study on LC/MS data preprocessing and analysis workflows based on the `xcms` Bioconductor package [158]. That is, the methods provided by the underlying libraries are not directly passed to the user level, but manually curated and wrapped into domain-specific accessible units of functionality. In his diploma thesis [334], Clemens von Musil developed a plugin for the automatic creation of SIBs from R packages based on their description files (which are similar to the manual pages of Unix operating systems). The plugin, called *jR*, accomplishes its task pretty well, and its use was demonstrated by means of different examples (matrix multiplication, lexical coverage, microarray data classification). However, the granularity of the provided functionality and thus of the workflows remains close to the corresponding R code, so that workflow design with jR-generated SIBs is too low-level for a framework that advocates working with coarse-grained functionality at the user level. Therefore jR has not been used for the integration of Bioconductor functionality into Bio-jETI.

6.2.2 Basic Microarray Data Analysis Pipeline

A basic, more or less minimal, example of a microarray analysis pipeline built with the SIB library described above is shown in Figure 6.6: The HGU95A spike-in benchmark data set is loaded and then preprocessed and filtered using the corresponding AffyExpress SIBs. After a differential expression analysis, the PubMed abstracts of articles potentially related to the top 10 differentially expressed genes are retrieved and saved. In this form, the analysis workflow corresponds clearly to the abstract microarray data analysis pipeline of Figure 6.5.

load spike-in   AffyExpress: AffyExpress: differential expression   get PubMed abstracts   save
example data   preprocessing   filtering   analysis   for top 10 genes   results

Fig. 6.6 Basic microarray data analysis workflow

6.2.3 Variable Microarray Data Analysis Pipeline

Figure 6.7 shows a variable workflow model that gives an impression of how the SIBs in the library can be combined into different microarray analysis pipelines. Similar to the variable the multiple sequence alignment workflow discussed in Section 3.2, the SIBs in the model are pre-configured to be readily executable, and the currently intended analysis steps can be included simply by redirecting branches. The boxes in the figure represent principal steps of microarray data analysis workflows, and contain different (combinations of) SIBs that realize corresponding tasks:

1. In the (naturally mandatory) input data loading step, the microarray raw data and the corresponding meta-data is loaded. It can be selected if one of the benchmark data sets is used that are readily available on the jETI server, or if the input data from the local file system is used.
2. Preprocessing is also mandatory. Here, `AffyExpressPreprocess` can be used to create an ExpressionSet object from the input data, or one of `RMA, GCRMA, Threestep` and `Express` followed by `CreateExpressionSet`.
3. It is recommended to apply one or more filtering steps to the expression values before applying further analyses.
4. Optionally, the expression values can be visualized in a HTML table (created by `Annaffy_aafTableInt` and then stored to the local file system).
5. Statistical analysis, for instance in order to identify the top differentially expressed genes, is then again considered mandatory. Optionally, a textual representation of the results can be written and stored to the local file system.
6. Finally, it is useful to retrieve further information about the top differentially expressed genes, for instance probe annotations (via `Annaffy_aafTableAnn`) or related PubMed articles (via `GetPubMedAbstracts`), and store the resulting (HTML) files.

The SIBs in the data loading, preprocessing, filtering, statistical analysis and annotation boxes that are highlighted by the light-gray box in the figure correspond to the basic analysis workflow described above. The workflow that is defined by the branches as shown in Figure 6.7 instead reads the input data from the local file systems, uses `Threestep` for the prepocessing and the `GenefilterKOverA` for expression value filtering. Then it creates and stores an HTML table from the expression values, before the differential expression analysis is carried out. Finally, an annotation table is created and stored.

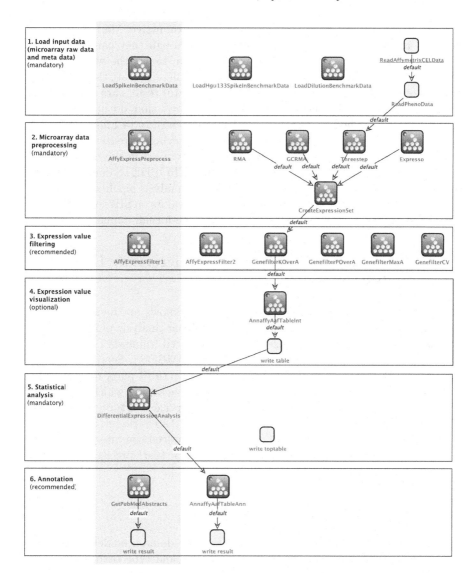

Fig. 6.7 Variable microarray data analysis workflow

Note that this variable model is by no means exhaustive. Many more variants are possible with the available SIBs, even though they can not be adequately represented by this kind of model. For instance, the workflow can be grow more or less arbitrarily when several different statistical analyses and annotations are included, or when (intermediate) results are stored more often. Hence, it is only natural to design workflows that do not fit in the simple schema implied by the abstract analysis workflow of Figure 6.6, and accordingly it is neither adequate nor expedient trying to represent the full variability explicitly in one workflow model. As detailed in the next section, the variability can be implicitly represented by a PROPHETS domain model, which can then be used to create various concrete analysis pipelines automatically. Moreover, also dependencies and other constraints on SIB combinations, which are not covered here, can be specified in the formal domain model.

6.3   Constraint-Driven Design of Microarray Data Analysis Workflows

In this section, the microarray data analysis workflows are revisited in the light of loose programming. Essentially being linear analysis pipelines, they can particularly well be synthesized by the presented framework. Like in the first two scenarios (phylogenetic analyses with EMBOSS, Section 3.3, and GeneFisher-P, Section 4.3), the services are based on third-party functionality (in this case from publicly available Bioconductor packages), which is however wrapped in order to be user-accessible. For the service and type taxonomies, self-defined vocabulary is used, as in the FiatFlux-P scenario (cf. Section 5.3). The reason here is that the EDAM ontology does indeed comprise vocabulary for microarray data analysis, but the terms are not (yet) precise enough for use in this application.

6.3.1   Domain Model

As for any PROPHETS workflow application, the domain model for the microarray data analysis scenario comprises service and type taxonomies, service descriptions and constraints. They are described in the following.

Taxonomies

Figures 6.8 and 6.9 show the service and type taxononmies, respectively, that have been set up for the microarray data analysis workflow scenario. The services are divided into specific *MicroarrayDataAnalysis* and *Common* services on the first level. The former distinguishes the analysis services further into *Preprocessing*, *Filtering*, *StatisticalAnalysis* and *Annotation* services, while the latter essentially categorizes further into *DataHandling* and *FileWriting* services. The type taxonomy is less comprehensive, essentially classifying the different *File* types directly as *HtmlFile*, *PdfFile* or *TextFile*, or as *MicroarrayData*, which covers *RawData*, *MetaData*, *AnalysisResults* and the different container data structures defined by Bioconductor.

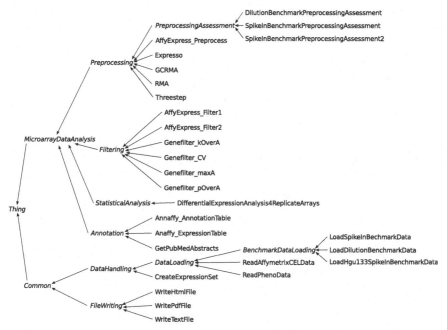

Fig. 6.8 Service taxonomy of the microarray data analysis domain model

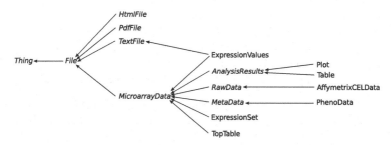

Fig. 6.9 Type taxonomy of the microarray data analysis domain model

Services

Table 6.1 lists the services in the domain model and their input/output data types. In addition to the specific Biocondcutor services already introduced in Section 6.2, some local services for the reading and writing of files and for the loading of benchmark data sets are contained in the domain model. Whereas the data loading/reading services require no inputs and the file writing services produce no outputs, all other services in the domain require one or more inputs and produce one or more outputs. Services that belong to the same (bottom-level) group of the service taxonomy often also have a similar input/output behavior. For instance, the *Filtering* services read and write ExpressionSets, while most of the *Preprocessing* services create ExpressionValues from AffymetrixCELData.

Constraints

The domain constraints in this example are used to describe the general overall structure of microarray data analysis pipelines, such as that *DataLoading*, *Preprocessing* and *StatisticalAnalysis* should be uses exactly once, and that as last step a *FileWriting* service should be used in order to have results stored to the local file system when the workflow finishes. Furthermore, domain constraints are used to express that the HTML-formatted results of *Annotation* steps should be stored immediately, and that the currently available differential expression analysis service is not suitable for application on the dilution benchmark data set. In terms of the template-based constraint editor of PROPHETS, the corresponding constraints are:

- Enforce the use of *DataLoading*.
  Enforce the use of *Preprocessing*.
  Enforce the use of *StatisticalAnalysis*.
- Do not use *BenchmarkDataLoading* more than once.
  Do not use *Preprocessing* more than once.
  Do not use *StatisticalAnalysis* more than once.
- Use *FileWriting* as last service in the solution.
- If *Annotation* is used, `WriteHtmlFile` has to be used next.
- If *StatisticalAnalysis* is used, do not use *Filtering* subsequently.
- At most one of `LoadDilutionBenchmarkData` and `DifferentialExpressionAnalysis4ReplicateArrays` may exist.

These constraints describe the substantial, common characteristics of microarray analysis pipelines built with this domain model. During workflow design, additional constraints can be used to guide the synthesis to particular solutions explicitly, for instance by demanding or avoiding the use of specific services.

6.3.2   Exemplary Workflow Composition Problem

Microarray data analysis workflows are typically pipelines (i.e., linear sequences of analysis steps), and thus linear synthesis is suitable to create the

Table 6.1 Services in the microarray data analysis domain model

Service	Input types	Output types
AffyExpress_Filter1	ExpressionSet	ExpressionSet
AffyExpress_Filter2	ExpressionSet	ExpressionSet
AffyExpress_Preprocess	AffymetrixCELData, PhenoData	ExpressionSet, PdfFile, Plot
Annaffy_AnnotationTable	TopTable	HtmlFile, Table
Annaffy_ExpressionTable	ExpressionSet	HtmlFile, Table
CreateExpressionSet	ExpressionValues, PhenoData	ExpressionSet
DifferentialExpressionAnalysis-4ReplicateArrays	ExpressionSet	TopTable, Table, Textfile
DilutionBenchmark-PreprocessingAssessment	ExpressionValues	PdfFile, Plot
Expresso	AffymetrixCELData	ExpressionValues
GCRMA	AffymetrixCELData	ExpressionValues
Genefilter_CV	ExpressionSet	ExpressionSet
Genefilter_kOverA	ExpressionSet	ExpressionSet
Genefilter_maxA	ExpressionSet	ExpressionSet
Genefilter_pOverA	ExpressionSet	ExpressionSet
GetPubMedAbstracts	TopTable	HtmlFile, Table
LoadDilutionBenchmarkData	-	AffymetrixCELData, PhenoData
LoadHgu133SpikeIn-BenchmarkData	-	AffymetrixCELData, PhenoData
LoadSpikeInBenchmarkData	-	AffymetrixCELData, PhenoData
ReadAffymetrixCELData	-	AffymetrixCELData
ReadPhenoData	-	PhenoData
RMA	AffymetrixCELData	ExpressionValues
SpikeInBenchmark-PreprocessingAssessment	ExpressionValues	PdfFile, Plot
SpikeInBenchmark-PreprocessingAssessment2	ExpressionValues	PdfFile, Plot
Threestep	AffymetrixCELData	ExpressionValues
WriteHtmlFile	*HtmlFile*	-
WritePdfFile	*PdfFile*	-
WriteTextFile	*TextFile*	-

whole analysis workflows automatically. Accordingly, the exemplary workflow composition problem for this scenario is not, in contrast to the previous
examples, embedded in some more or less complex workflow skeleton, but
conceived for the start-to-end synthesis of the analysis pipeline. As Figure
6.10 (top) shows, such a minimal loosely specified model for synthesis with
PROPHETS consists of two (Noop) SIBs (here simply labeled start and
end) that are connected by a loose branch.

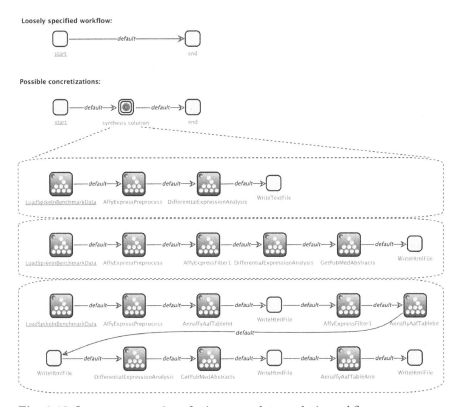

Fig. 6.10 Loose programming of microarray data analysis workflows

The bottom of the figure shows three of the numerous possible concretizations of the loose branch. The first example synthesis solution is one of the
shortest possible solutions, consisting merely of a *BenchmarkDataLoading*,
a *Preprocessing*, a *StatisticalAnalysis* and a *WriteFile* step. As it is recommended to filter the preprocessed data prior to applying statistical analyses,
an additional constraint that enforces the use of a *Filtering* services may
be added at workflow design time. Together with a constraint that enforces
the use of an (also recommended) *Annotation* step, the synthesis solutions
will comprise at least two additional steps, as shown in the second example:

`AffyExpressFilter1` and `GetPubMedAbstracts` are inserted, and accordingly the `WriteTextFile` step at the end is replaced by a `WriteHtmlFile` step that stores the list of retrieved PubMed abstracts. However, naturally also more comprehensive solutions are possible with this set of constraints: the third example in the figure shows a solution similar to the second one, where in addition the expression values are visualized before and after the preprocessing step, and where finally an annotation table is created for the top differentially expressed genes.

In fact, the essential general characteristics of micorarray analysis pipelines are already covered by the defined domain constraints: When applying no constraints at all, the search for synthesis solutions exceeds the 1,000,000-solutions default when searching for solutions in search depths greater than 7, while with the domain constraints 1,779 solutions of length less or equal than 7 are found. As already shown in the previous chapters for the other three scenarios, especially for the loose programming of phylogenetic analysis workflows in Section 3.3, the user can influence the synthesis results easily by varying the constraints at workflow design time. Particularly effective is the expression of concrete intents, for instance by explicitly demanding the use of a specific data loading or annotation service. For example, the additional enforcement of `LoadSpikeInBenchmarkData` and `GetPubMedAbstracts` leaves 85 solutions that are found in search depth 7.

Part III

Discussion

7

Lessons Learned

The application examples presented in Part II (Chapters 3 – 6) demonstrate that in silico experiments of different flavors can be realized with the Bio-jETI framework. They cover a broad range of typical bioinformatics work-flow scenarios, concerned with different thematic areas, different software components, different service technologies, and also workflows of different complexity. They illustrate how the constraint-driven workflow develop-ment methodology helps mastering the manifold workflow variants and how PROPHETS' ability to flexibly formulate domain-specific and problem-specific constraints supports the workflow development process. As such, the application scenarios provide on the one hand qualitative evidence of the method's applicability on real-world application scenarios, and on the other hand they are suitable as a proper basis for further considerations about the applied workflow development technology. In this context, this chapter focuses specifically on the "lessons learned" regarding the constraint-driven workflow design methodology that has been introduced to the Bio-jETI frame-work in the scope of the work underlying this book.

Section 7.1 discusses the evaluation of particular key figures of the synthe-sis process in order to quantify the capabilities and limitations of the synthesis framework systematically. In fact, the applied evaluation scheme is suitable to show how effectively the constraint-driven method can guide the automated search for workflows according to the user's (high-level) intents, but that its performance is impacted by the state explosion of the search space that the synthesis algorithm faces sooner or later. Interestingly, however, constraints are also a key means with regard to dealing with the state explosion problem: They reduce the size of the search space as they narrow the set of admissible solutions. Thus, they do not only guide the search towards the actually in-tended solutions, but at the same time help to delay performance problems caused by state explosion.

Section 7.2 then summarizes the experiences gained during working on the application scenarios and the results from the evaluation. The resulting *loose programming pragmatics* provide guidance for domain preparation and service integration for workflow applications in general, and for semantic domain modeling and synthesis application the scope of the loose programming framework in particular.

7.1 Synthesis Evaluation

Before starting with the details of the evaluation, the following briefly summarizes and compares the four considered application scenarios with particular regard to the constraint-driven workflow design with PROPHETS:

Application Scenario 1: Phylogenetic Analysis Workflows

The EMBOSS domain model comprises the largest amount of services (more than 430) among the four examples. It has been automatically generated from the ACD files provided with the EMBOSS release, which also contain semantic annotations in terms of the EDAM ontology. This application is particularly challenging for the synthesis method, since the domain contains many services for similar purposes (sequence analysis) that also work on similar data types, so that the synthesis can easily find extremely many solutions already in relatively small search depths. Consequently, the provisioning of adequate domain-specific and problem-specific constraints is essential in order to arrive at a manageable amount of solutions.

Application Scenario 2: GeneFisher-P

The GeneFisher-P domain model also applies the EDAM terminology, but the service annotations themselves have been defined manually. It is much smaller than the EMBOSS domain (only comprising 10 services) and also much more specific (tailored to primer design workflows). Hence considerably less workflows are possible based on the GeneFisher-P domain model and thus the set of solutions that is obtained by the synthesis is more or less manageable also without defining comprehensive sets of additional constraints. The exemplary workflow composition problem also illustrates the over-constraining of specifications and some of the problems that can be caused by imprecise service interface descriptions.

Application Scenario 3: FiatFlux-P

The FiatFlux-P domain model does not apply EDAM terminology, since the required terms are not (yet) covered by the ontology. Accordingly, service annotations and taxonomies have been created manually and tailored to the specific application. The FiatFlux-P domain is only slightly larger than the GeneFisher-P domain (12 services), so that the number of solutions that have to be handled is also within a feasible range. As it contains a handful of services for the creation of data "from scratch", a couple of constraints are required to prevent their unrestrained use by the synthesis. Some specific constraints of the FiatFlux-P application can, however, not be expressed with the constraint templates provided by the PROPHETS plugin and require the definition of additional temporal logic formulae.

Application Scenario 4: Microarray Data Analysis Workflows

The domain model for the microarray data analysis workflow scenario comprises 27 services and is thus considerably larger than the previous two example domains. Like FiatFlux-P, it does not apply the EDAM terminology, however because the microarray-related terms in the ontology are not precise enough for describing the services and types in this application scenario. As microarray data analyses are typically linear sequences of services ("pipelines"), the workflows for this example can be entirely created by PROPHETS, which is based on a linear-time logic synthesis algorithm. Furthermore, the workflows of this example are comparatively long, typically comprising at least 6 principal analysis steps, which are often themselves composed of different services. Accordingly, creating proper microarray data analysis pipelines manually is not that easy, and their (semi-) automatic synthesis achieves a considerable simplification.

As discussed in greater detail in [233], the adequacy of the synthesis solutions largely depends on the domain model: the closer a domain model is tailored to a specific application, the less effort (in terms of formulating additional constraints) is required from the user to arrive at the actually intended workflows. In this regard, considerable differences can also be observed between the example applications:

- The EMBOSS domain model contains a large number of tools for similar purposes (sequence analysis), but has not been designed for a particular application. Consequently, it provides more possibilities for service combination than a user can handle, and additional knowledge has to be incorporated in order to constrain the solution space to a manageable size. In addition, the EMBOSS example shows that service interface annotation in terms of a comprehensive domain vocabulary (EDAM) alone is not sufficient.
- In contrast, the manually defined GeneFisher-P and FiatFlux-P domain models have been tailored to a specific application. That is, the domain vocabulary has been designed specifically, the services provide exactly the functionality that is needed, and additional domain-specific knowledge has already been formalized and integrated into the domain model. This shows that in domains with many purpose-specific services, a good amount of the domain knowledge is often implicitly integrated in the service interface design, and does not have to be provided via constraints explicitly.

- The microarray data analysis example is in this sense in between: its domain model is more specialized than that of the EMBOSS scenario, but by far not as concise as those of GeneFisher-P and FiatFlux-P. Hence constraints are essential to describe the intended overall workflow structure adequately, and to express specific dependencies between services as well as to communicate concrete user intents.

The descriptions of the constraint-driven workflow design applications in Part II contain only selected numbers for illustrating the effects of individual constraints. This section illustrates the impact of the constraints more expressively, discussing in greater detail how the domain model in general and the constraints in particular influence the synthesis solutions and the synthesis performance. In particular, the following key numbers provide a proper basis for the systematic evaluation of the impact of different constraint combinations:

- *Search depth*:
  To explore the actual solutions, the synthesis framework performs a breadth-first search. That is, when a particular search depth d has been processed, all solutions of length smaller or equal to d have been found.
- *Number of solutions*:
  The synthesis framework does not only provide the total number of solutions that is available after the execution of the algorithm has finished, but also traces the number of solutions that are found in each individual search depth. This allows for a very precise monitoring of constraint effects over several search depths. Note that the largest number of solutions is obtained when no constraints are applied, as then the number of solutions directly corresponds to the number of solutions that are implied by the synthesis universe alone. With constraints, less or equally many solutions are found.
- *Number of visited nodes*:
  Analogous to the monitoring of the number of solutions, the synthesis framework traces the total and the depth-wise numbers of nodes in the synthesis universe that are expanded by the algorithm during the search for solutions. The number of visited nodes is not necessarily proportional to the number of solutions that are found, but determines the runtime performance of the search.

According to this, Sections 7.1.1 and 7.1.2 systematically compare the numbers of solutions and visited nodes, respectively, for several search depths and for different combinations of constraints. To ensure comparability of the results, all experiments for obtaining the numbers were carried out on the same machine (equipped with a – comparatively fast – Intel Core i5 2500k CPU with 3.3 GHz and 8192 MB RAM) and with the same version of BiojETI (based on jABC 3.8.5 and PROPHETS 1.2 as released in September 2011).

7.1.1   Solutions

For assessing the impact of the constraints on the (number of) synthesis solutions, the numbers of synthesis solutions for the different constraint combination are plotted for the individual search depths. By this means Figures 7.1, 7.2, 7.3 and 7.4 in this section show that the exponential growth of solutions that is typically observed for unconstrained syntheses can effectively be restrained through the application of constraints that concisely express the intended workflow structure.

The main reasons for the exponential increase of solutions are:

- *Similar services*:
  The synthesis algorithm generates sequences of services based on the compatibility of their input and output types. Accordingly, if no constraints are provided, all sequences that are technically possible are returned. However, also in the presence of additional constraints, the services' input/output specification is the main source of information for their combination. Most application domains, and in particular the domain models for the bioinformatics scenarios considered in this book, contain many similar services in the sense that they work on similar input types, produce similar output types, or even simply modify a data item so that the input and output data types are identical. Obviously, this provides extremely many possibilities for their combination.
- *Accumulation of data:*
  Most services in the considered application domains generate new data as they process their inputs to produce their outputs. Accordingly, the amount of available data (types) increases with the length of the solution, offering more and more possibilities which services to use.
- *Infinite behavior:*
  Often, individual services or sequences of services can be used repeatedly, constituting loops that can be repeated infinitely often. Consequently, it is usually impossible to explore the solution space completely, and the domain models typically comprise the possibility of infinite workflows. Although it is principally possible to apply constraints that explicitly suppress the repeated use of (individual) services, this is not adequate for all services in general.

Figure 7.1 summarizes the development of the numbers of solutions for the different sets of constraints applied in the EMBOSS scenario discussed in Section 3.3. The left plot gives the numbers in normal scale, the plot at the right contains the same number in logarithmic scale for better discrimination of the smaller numbers. The legend within the left plot shows which lines in the plots correspond to which sets of constraints.

As already discussed in Section 3.3, in the unconstrained case the maximum capacity of the PROPHETS solution store used in the applications (1,000,000 solutions) is reached in a search depth of 4. Accordingly, the plot

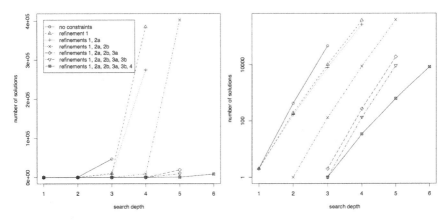

Fig. 7.1 Synthesis statistics (number of solutions) for the EMBOSS scenario

for this case runs only until a search depth of 3, where the last complete numbers have been obtained. Similarly, several lines run only until a search depth of 4 or 5, which means that for these cases the maximum capacity of the solution store has been reached in depth 5 or 6, respectively. Still, these lines show the effect of the constraints on the number of solutions, enabling the synthesis algorithm to proceed to greater search depths. In the last case the solutions of length 6 can be fully explored, resulting in 8,317 remaining solutions.

In total, the application of constraints and the use of particular synthesis configurations tames the growth of solutions effectively: instead of exceeding the solution store capacity already in a search depth of 4, the synthesis finally returns a comparatively handy set of solutions of length less than or equal to 6. Nevertheless the reached search depth is relatively small. Although further constraints could restrain the solutions further and thus avoid exceeding the capacity of the solution store also for greater search depths, eventually the execution time of the synthesis exceeds feasible ranges. Consequently, when working on the discussed application scenarios, the synthesis was simply aborted when it became clear that it would be running "too long", which typically meant running for days or weeks and not for minutes or hours any more. (The next section discusses the performance of the synthesis in greater detail.)

Figure 7.2 visualizes until a search depth of 12 the number of synthesis solutions that are obtained in the GeneFisher-P scenario (cf. Section 4.3) when no constraints, the domain constraints, and constraints for the enforcement/avoidance of *DNA backtranslation* and *Sequence alignment* are applied. In the unconstrained case, the number of solutions grows exponentially, returning 101,842 solutions in depth 12. The domain constraints reduce the number of solutions in depth 12 to 1,086 solutions, and additionally cause that no solutions are found before a search depth of 5. Enforcing the use of *DNA backtranslation* and *Sequence alignment* services yields that the shortest two solutions have a length

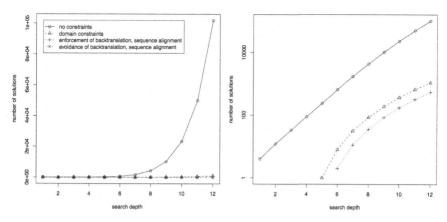

Fig. 7.2 Synthesis statistics (solutions) for the GeneFisher-P scenario

of 6, while the number of solutions in depth 12 is 546. As already discussed in Section 4.3, constraints that avoid the use of *DNA backtranslation* and *Sequence alignment* over-constrain the synthesis specification and no results are returned.

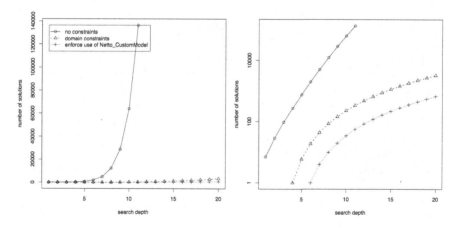

Fig. 7.3 Synthesis statistics (solutions) for the FiatFlux-P scenario

As Figure 7.3 shows, the unconstrained synthesis of FiatFlux-P workflows as described in Section 5.3 leads to unmanageable amounts of solutions very fast: 277 solutions are obtained in a search depth of 4, already 136,578 in depth 11, and the capacity of the solution store is reached when exploring search depth 13 (not shown in the figure). The domain constraints alone tame the growth of solutions very effectively, causing the synthesis to return

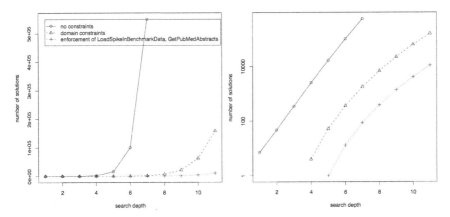

Fig. 7.4 Synthesis statistics (solutions) for the microarray data analysis scenario

a single solution of length 4 and 344 solutions of length less than or equal to 11. Even in a search depth of 20, a comparatively small number of 3,281 solutions is returned. An additional constraint that enforces the use of a particular analysis step (Netto_CustomModel) even leads to 0 (depth 4), 56 (depth 11) and 680 (depth 20) obtained results.

When synthesizing microarray data analysis workflows in the scope of the scenario described in Section 6.3, the maximum capacity of the solution store is reached in a search depth of 8 if no constraints are applied. Due to the complexity of the analysis, there are, however, several workflow variants that involve more steps, so it is clearly necessary to enable the synthesis to proceed to greater search depths. Figure 7.4 shows that already the domain constraints restrain the growth of the amount of solutions effectively, so that 6,855 results are returned in a search depth of 8. Additionally enforcing the use of a particular benchmark data set (LoadSpikeInBenchmarkData) and the GetPubMedAbstracts annotation service tames the amount of obtained solutions further, so that 385 solutions of length less than or equal to 8 remain.

7.1.2   Performance

The runtime performance of the synthesis algorithm largely depends on the domain model: The synthesis universe (cf. Section 2.3.1) is constituted by the (static) service descriptions, more precisely by the directed graph constructed from their behavioral characterization in terms of input and output data types. Thus, the number of states in the synthesis universe is exponential in the number of services in the domain model, an effect that is commonly known as *state explosion* [324]. In other words, the more behaviorally similar tools (i.e., tools with similar inputs or outputs) are available in the domain model, the more possibilities exist for their combination into workflows and the greater is the solution space that has to be searched.

As detailed in Section 2.3.1, the synthesis universe is too large to be generated explicitly, and hence the required nodes are expanded on the fly during the search for solutions. The overall execution time of the synthesis is in fact clearly dominated by the execution time of the search, which corresponds directly to the number of expanded nodes. Thus, the evaluation of the applications shows that the synthesis performance has to be improved in order to allow for truly seamless and smooth workflow design: several minutes can already be needed for synthesizing sequences consisting only of a handful of services. To give an impression: running on the machine that has been used for carrying out the experiments for the evaluation, the algorithm is able to expand about 250,000 nodes per second. There, the search for solutions for the (comparatively small) synthesis problems discussed for the EMBOSS scenario in Section 3.3, for instance, runs for up to 3 minutes when considering solutions until a length of 4, and already up to 257 minutes (i.e. about 4.5 hours) when searching until a depth of 5. Note that although the EMBOSS domain model with its about 430 services is the largest among the presented examples, is still far from future real-world domain models that would have to cover, for instance, whole service repositories like the BioCatalogue [112], which at the time of this writing comprises almost 2300 services.

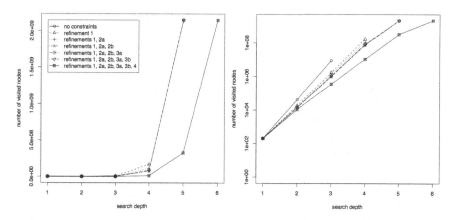

Fig. 7.5 Synthesis statistics (number of visited nodes) for the EMBOSS scenario

Figure 7.5 illustrates the number of visited nodes and thus the runtime performance of the synthesis in the EMBOSS scenario in greater detail. Obviously, constraints can reduce the size of the search space as they decrease the number of solutions, and thus speed up the search for solutions: For instance, using only the first constraint leads to 168,239,054 visited nodes in depth 4, while 10,847,627 nodes are visited when applying all constraints,

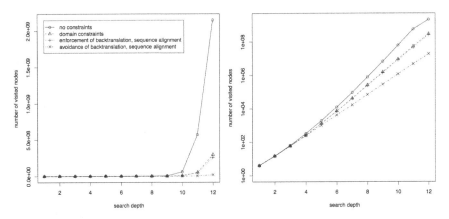

Fig. 7.6 Synthesis statistics (visited nodes) for the GeneFisher-P scenario

which can be about 7 times faster (more precisely: about 30 seconds instead of roughly 3.5 minutes on the machine on which the experiments were carried out). As another example, the synthesis expands 3,642,130,982 nodes in depth 5 when applying constraints 1 and 2, while the application of all constraints leads to 327,212,871 visited nodes, which is still about 6 times faster (about 14 minutes instead of 82). However, the constraints reduce the number of visited nodes not always as clearly as they restrain the number of solutions. For example, the number of visited nodes in depth 5 for the refinements 1–3 are the same order of magnitude (roughly 3.5 billion expanded nodes), whereas the corresponding numbers of solutions are clearly different (cf. Figure 7.1).

The EMBOSS domain model is indeed a prominent example for a large collection of similar services that induce a high branching factor in the synthesis universe and accordingly a large solution space. Consequently, the search depths that can be reached in acceptable synthesis execution times are comparatively small unless extremely concise constraints are provided. In contrast, the GeneFisher-P and FiatFlux-P domain models comprise considerably less and furthermore behaviorally distinct services, and hence the corresponding solution spaces are smaller. As Figures 7.6 and 7.7 show, there greater search depths can more easily be reached, as the number of visited nodes "explodes" much later.

Figure 7.6 shows that the number of visited nodes in the GeneFisher-P scenario is more or less equal and comparatively small in the beginning, regardless of the actual constraints that are applied. In a search depth of 10, the difference becomes more striking, as 60,929,232 nodes are visited in the unconstrained case, whereas the three constrained cases cause the synthesis algorithm to expand between 1,165,825 and 9,091,818 nodes. This took between 2.8 and 17.5 seconds instead of 1.5 minutes. In depth 12, the unconstrained synthesis explores a total of 5,482,162,778 nodes, while the most stringently constrained case visits 19,048,121 nodes, which has been roughly 200 times faster (more precisely: about 45 seconds instead of about 2.5 hours).

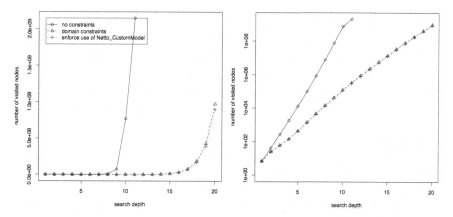

Fig. 7.7 Synthesis statistics (visited nodes) for the FiatFlux-P scenario

The results for the FiatFlux-P scenario shown in Figure 7.7 are principally similar, only that even a search depth of 20 can still be reached within acceptable execution times. In the unconstrained case 7,963,870,317 nodes are expanded in search depth 11, which means that depth 12 can not be explored within acceptable execution times. The domain constraints and the additional enforcement of a particular service effectively tame the number of visited nodes, so that an increase of the execution time is not noticeable before a search depth of 15, where 14,257,817 and 12,967,147 nodes are visited

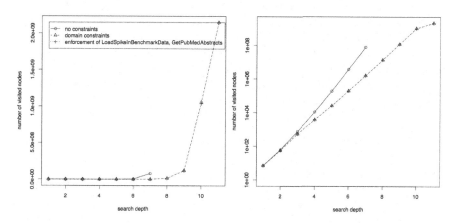

Fig. 7.8 Synthesis statistics (visited nodes) for the microarray scenario

for the two constrained cases, respectively. Now the number of visited nodes per search depth increases faster, until 973,788,715 and 894,577,045 nodes are expanded in a search depth of 20, which still took less than one hour.

Finally, Figure 7.8 confirms the expectation that also the number of visited nodes in the microarray data analysis scenario can be influenced by the application of constraints. However, as the domain model is also more complex and enables a greater variability than those of the GeneFisher-P and FiatFlux-P scenarios, the decrease is not as striking as in the previous two examples and the search depth that has finally been reached is not as high. In fact, the domain constraints and the additional enforcement of a particular data provisioning service and a particular annotation service lead to roughly the same numbers of visited nodes, namely 9,320,707,579 and 9,148,899,935 in depth 11, respectively. Exploring them took nearly 10 hours.

The evaluation of the number of visited nodes shows that constraints can decrease their number effectively and thus decrease the time that is required for exploring a particular search depth. At the downside, the more constraints are applied, the longer takes the construction of the automaton from the constraints prior to the actual search. However, this step is usually faster and in summary not crucial for the runtime performance of the complete synthesis process.

7.2   Loose Programming Pragmatics

In the light of complex and evolving applications, it is impossible to provide a semantic workflow environment that suits all application scenarios equally well. In particular, there is no such thing as a unique "perfect" domain model, just as there is no ultimate synthesis strategy that fits all situations. However, the experiences gained from working on the applications revealed some rules-of-thumb for domain modeling and synthesis application that help to approach loose programming pragmatically and exploit its capabilities as far as possible. In the following, Section 7.2.1 discusses principles for the integration of services for workflow applications and as adequate foundations for semantic domain models, before Section 7.2.2 focuses on pragmatics for the actual semantic domain modeling and Section 7.2.3 presents strategies for the pragmatic use of workflow synthesis.

7.2.1   Service Integration

Clearly, the services in a domain model should in the first place provide the *functionality* that is required for realizing the envisaged application(s). However, the *granularity* of the services in the domain model is an equally central and crucial aspect, especially when working towards user-level workflow design (cf., e.g., [206, 197, 261, 211, 199]).

As detailed in the introduction (Section 1.1.1), services are basically readily executable "nuggets of functionality" that are ready for use in complex applications. More specifically, services as understood in the scope of this work are clearly defined units of functionality that abstract from the interfaces of the underlying implementations in order to provide the functionality at an application-specific, less technical, and hence user-accessible level. Ideally, the specification and design of services is approached in a top-down and use-case-oriented fashion, identifying the requirements from the user's or application's perspective. A frequent problem with existing services, however, is that they are usually conceived from a programmers' perspective, which often differs considerably from that of the user: While programmers tend to think in technical details of the concrete service interface implementations, users typically think in the pieces of functionality required for their applications. As a consequence, service interfaces are often convenient to use for application programmers, but not for users of workflow environments, who want to work on a more abstract level.

In fact, many existing tools and services provide useful functionality, but are not directly adequate for use in workflow applications due to inappropriate interfaces. Thus, turning them into adequate SIB libraries may require additional service conception efforts. More precisely, either a service of adequate granularity is *directly* usable, or the SIB developer can *combine* different small steps into an adequate larger one, or also *split* a large step into adequate smaller ones. The following explains and illustrates these three principal possibilities in greater detail. Subsequently it is discussed that the same principles also apply for the data types that are associated with the services.

Directly Using Services

Ideally, the workflow building blocks for a particular application scenario can directly be derived from existing services. As already sketched above, however, existing services are often simply not conceived for use in user-oriented workflow environments like Bio-jETI, and are thus not directly usable.

Interestingly, it has often been possible to use "classical" command line tools more or less directly, while for many "modern" (esp. Web Service) remote interfaces straightforward integration did not result in user-level SIBs. In fact, command line tools are typically designed to execute particular, well-defined tasks, and usually all inputs and configuration options can be provided upon invocation, so that their execution runs completely autonomous ("headless"). Furthermore, command line tools typically work on files, which are per se more user-level than the programming language entities (such as Java objects) that are required for the communication with, e.g., Web Service APIs. Accordingly, jETI services, which are designed to provide convenient (remote) access to file-based command line tools, are inherently closer to the user level than web services.

Among the application examples, this has become most evident for the phylogenetics analysis workflow scenario discussed in Chapter 3: The first workflow examples in this scenario (cf. Section 3.2) are based on the DDBJ web services, which provide access to several standard sequence databases and sequence analysis tools. Technically, the DDBJ web services are a (positive) exception among bioinformatics web services, as they provide extremely simple service interfaces and are hence convenient to use for workflow applications. However, as they provide only a limited range of functionality, the later, more comprehensive workflow examples of the phylogenetics scenario are based on the EMBOSS tool suite. Principally, most EMBOSS tools are readily available as web services, provided by the EBI's Soaplab project [14, 15]. However, as many EBI web services, they provide rather complicated interfaces (introducing loads of technical data types and requiring several service calls for executing one tool, as the services are run asynchronously) and are hence inconvenient to use. Consequently, the EMBOSS tools have simply been provided as jETI services. Due to the availability of complete command line interface descriptions in the ACD files (cf. Section 3.3), this tool integration process could even be completely automated. Also the other three application scenarios described in this book apply the jETI technology for service provisioning.

Combining Services

Unless the adequate service interface corresponds directly to a single call to a tool or API, this typically involves proper combination of several small steps into a single, more abstract unit of functionality that then constitutes the resulting service.

In the course of investigating further application areas of the Bio-jETI framework as described in this book, there have been different attempts to provide systematic integrations of specific programming languages or APIs. For instance, the *jR* plugin for the jABC framework, developed by Clemens von Musil in his diploma thesis [334], provides means for automatic generation of SIBs for GNU R library functions and also for the execution of the corresponding workflow via a (remote) R engine. The integration and execution facilities work properly, but as the granularity of the generated SIBs essentially corresponds to single R statements, they are at the end of the day simply too low-level for the application experts who are the actual target users of the framework. Analogous experiences have been reported from similar experiments concerning a Matlab integration (Wolfgang Jansen, Potsdam University, personal communication).

Consequently, for the integration of such functionality it is more adequate to proceed as in the microarray data analysis scenario described in Chapter 6. There, the R (or, more precisely, Bioconductor) functions that constitute the "heart" of the envisaged services, are wrapped appropriately. Concretely, the service does not call the library function directly, but a small script that also executes the smaller steps that are required for preparing the actual library call (such as the assembly of specific input data types, loading of dependencies, technical configurations etc.) and possibly also for postprocessing of the result of the library call before it is returned to the client. Following this pattern, a SIB can in fact easily provide a user-accessible unit of library functionality, as has been shown for the microarray data analysis scenario.

Splitting Services

In contrast, adequate services can also result from splitting a large unit of functionality into several smaller ones. This has been the case, for instance, for the GeneFisher-P and FiatFlux-P workflow scenarios (cf. Chapters 4 and 5, respectively), where initially monolithic applications have been decomposed into their principal, exchangeable steps in order to provide the services for more flexible, workflow-based implementations of the original applications.

As the GeneFisher web application had itself already used web services for particular steps of the primer design process, these were directly re-used as services for GeneFisher-P. The other required functionality was mainly available from the jABC's Common SIB library, so that the service provisioning for this application scenario was comparatively straightforward. For FiatFlux-P, in contrast, the underlying FiatFlux software had to be adapted prior to being usable for service integration, and accordingly there were no previously defined service interfaces available. In fact, the services that had been defined first soon turned out to be too fine-grained and were accordingly merged again into the larger, principal analysis steps as described in Chapter 5.

Adequate Data Types

For the data types associated with the services, generally the same integration principles as for the services apply, likewise following the paradigm that a SIB can (and should) hide everything from the user that is not relevant to him. Accordingly, data types of the underlying tool or service that are considered user-accessible can be directly used in the SIB that integrates the corresponding functionality. However, the application-level data types of the SIB do not have to be identical to the technical data types of the underlying service implementation. In particular, data that is not relevant from the user's perspective should not appear in the SIB interface, but simply handled internally. Similarly, if the original data type is too complex, it can simply be split and only the application-relevant parts be exposed as SIB parameters.

A direct mapping of service parameters to SIB parameters has been implemented, for instance, for the DDBJ SIBs of the phylogentics analysis workflow scenario (cf. Section 3.2.1. These web services do in fact simply use standard data types for the representation of bioinformatics entities, such as strings for molecular sequences and integers for amounts. Hence, the parameters defined by the web service interface can easily be used as SIB parameters. As contrasting examples, the EBI and BiBiServ web services considered for the phylogenetic analyisis and GeneFisher-P scenarios (cf. Chapters 4 and 5) use custom data types in their interfaces to combine all input data into a single input object. These are indeed easily accessible for a programmer, but not for a non-technical user. Hence, the corresponding SIBs expose only the application-relevant data as parameters, and handle the assembly and evaluation of the service's input/output objects internally. The complex data structures of the FiatFlux-P and microarray data analysis scenarios (cf. Chapters 5 and 6, respectively) are completely kept within the raw data files and those that are produced by the different services. Accordingly, the SIB parameters in these applications contain references to the files and additional configuration information, but no actual data.

7.2.2   Semantic Domain Modeling

As detailed in Section 2.3.2, the semantic domain models of PROPHETS are entirely based on symbolic names for services and data types, and the semantic service descriptions are completely decoupled from the SIBs that implement the actual functionality. The latter is one of the central improvements in contrast to earlier implementations of the synthesis method, where the semantic interface annotations had to be provided by the SIB programmers already during SIB implementation and could not be changed later. Clearly, like this the semantic description(s) provided for a service can be freely defined by the domain modeler: he can, for instance, use his own terminology, use the same service for different purposes, or simply omit unnecessary details in the interface description.

Nevertheless, domain modeling remains a challenging task, which has to take into account a plethora of aspects and naturally depends massively on the different characteristics of the various application domains. Accordingly, each domain modeling process is different, and not too many general rules can be laid down. As in all software engineering processes, finally a good amount of experience is required in order to determine the adequate levels of abstraction for components and interfaces. However, two general domain modeling paradigms can be identified:

1. define a *precise domain vocabulary*, and
2. keep the semantic *service interface descriptions simple*.

What this means in the context of the Bio-jETI framework is discussed in greater detail in the following, accompanied by examples from the considered application scenarios and other Bio-jETI and jABC applications.

Note that since also the most carefully designed domain model can not be expected to suit all possible application scenarios equally well, loose programming explicitly encourages the workflow designer to "play" with the semantic domain model, which is not irrefutable after the initial domain modeling process: The workflow designer is free to adapt the domain model according to his concrete needs and bring in his specific expert knowledge at any time. This is in fact specifically supported by the PROPHETS plugin: the user can change the service interface descriptions, edit the service and data type taxonomies, and remove and add constraints via the graphical user interface. As detailed in Section 8.1.4, this extremely flexible domain modeling approach is so far unique among the frameworks for automatic workflow composition in the bioinformatics domain.

Precise Domain Vocabulary

For the semantic service interface descriptions, a domain vocabulary is required that provides a precise terminology to describe and clearly distinguish the services and data types in the domain. Especially in scientific applications, the data that is processed by the services in the workflows is central, and accordingly it is important that the domain vocabulary comprises domain-specific terminology for referring to data. Since in the loose programming framework the data types in the semantic domain model are symbolic and completely decoupled from the technical data types of the underlying SIBs/services, the domain modeler can in fact deliberately define the data types that he considers adequate for the synthesis application.

Clearly, existing ontologies or taxonomies should be reused (if available). The currently most popular example of a readily available (technical) ontology in the bioinformatics domain is EDAM (cf. Section 3.3.1), which provides a comprehensive vocabulary for the annotation of all kinds of bioinformatics entities. However, if the existing vocabularies are not precise enough, they should simply be extended or replaced by more concise terminology. As detailed in the following, the definition of the domain vocabulary varied considerably between the four application scenarios of this book.

Owing to community efforts, the EMBOSS tool suite, which forms the basis for the first loose programming application scenario presented in Chapter 3, has already been completely annotated in terms of the EDAM ontology. Thus, the service and type taxonomies of the respective PROPHETS domain model could simply be derived from the corresponding branches of the EDAM ontology, and the services from the EMBOSS suite could be characterized and sorted into the taxonomies according to the available EDAM annotations. As demonstrated in detail in Section 3.3, this domain model provides a proper basis for effective constraint-driven synthesis of phylogenetic analysis workflows.

For GeneFisher-P (the workflow scenario discussed in Chapter 4), which is also concerned with several sequence analysis steps, it was again possible to use the EDAM ontology as basis for the taxonomies in the domain model. Semantic categories for the involved computational services (such as *PCR primer design* for the GeneFisher service) as well as for the associated data types (such as *PCR primers*) are contained in EDAM. In this scenario, there is furthermore a small number of services (providing common functionality such as file reading and writing) whose semantic characterizations are not specifically covered by the EDAM ontology. Usually, grouping too many services within too generic semantic categories comes with the risk of preventing the synthesis framework to distinguish them properly. However, as the domain model here is comparatively small, and the services in question are quite different per se, it turned out to be sufficient to use the closest, yet quite generic, EDAM terms for their annotation. Thus, also in the GeneFisher-P workflow scenario the taxonomies of the domain model are based on an existing ontology.

In contrast, the domain models for the FiatFlux-P and microarray data analysis workflow scenarios do not apply EDAM terminology. Although there are terms for the fields of metabolic flux analysis and microarray data processing contained in the ontology, they are very generic and as such not adequate for the annotation of services that perform very particular operations. Hence, own terminology has been defined for these domain models, specifically providing the semantic categories that are relevant for the respective applications at an adequate level of abstraction. For instance, the service taxonomy of the FiatFlux-P application defines abstract categories for *DataInput*, *DataOutput*, the actual *FiatFlux* services and additional *Helpers*, some of these with further sub-categories. Similarly, in the microarray data analysis scenario the domain model defines service categories for the different domain-specific *MicroarrayDataAnalysis* services, as well as for common functionality such as *DataHandling* and *FileWriting*, and further, more specific sub-classes of these. In both applications, also the defined data types are semantically classified accordingly.

Simple Service Interface Descriptions

The provisioning of adequate services in a domain model is already relevant if the services are to be composed by a human user. As detailed above, in particular services of adequate granularity are needed as foundation for a good domain model. In order to make effective use of a service, already a human user has furthermore to know as exactly as possible what the services does. Naturally, proper service provisioning also involves the inclusion of adequate service documentation, so that the user can easily find all relevant information about the service. Automatic workflow composition methods like the synthesis algorithm of the loose programming framework require even more precise semantic service descriptions, that is, such that are formally defined and programmatically accessible. Technically, the required functionality for this task is readily provided by the framework, but the actual challenge for the domain modeler is to define adequate service interface descriptions.

As a general paradigm, service interfaces should always be *as simple as possible*. More precisely, services should be "atomic" in the sense of providing a clearly defined, basic unit of functionality with deterministic input/output semantics. Semantically complex services, for instance such with alternative runtime behavior under different circumstances (*OR-semantics*), can not be described and handled properly by the synthesis framework.

For example, consider the ClustalW service that is used in the GeneFisher-P scenario (cf. Chapter 4, respectively). Using a set of molecular sequences as input, it computes a multiple sequence alignment. More precisely, if the input is a set of nucleic acid sequences, it computes a nucleic acid alignment, and if the input is a set of amino acid sequences, the result is an amino acid alignment. This alternative behavior according to the concrete nature of the input can as such not be expressed in the synthesis framework. As this distinction between the sequence types is not required in the scope of the considered workflow scenario, however, there ClustalW is simply annotated using a *Sequence* as input and producing a *Sequence alignment (multiple)* as output (cf. Section 4.3.1).

Generally, however, treating services as multi-purpose services that can be applied to different input data types often leads to a loss of precision, as the output is typically described at an equally abstract level. That is, the input data is "lifted" to a higher level from the perspective of the synthesis algorithm. As a consequence, actually matching services are not recognized any more. For example, the result of ClustalW as defined above is a (general) multiple sequence alignment, regardless whether the input sequences were nucleic acid or amino acid sequences. Hence, the synthesis is not able to recognize the possibility to use the result as input for a special service that works on multiple alignments of a particular sequence type, such as, for instance, the `fdnaml` and `fproml` phylogenetic tree construction services from the EMBOSS scenario (cf. Section 3.3.2).

An interesting feature of PROPHETS domain models in this regard is the possibility to define multiple service interface descriptions for one and the same underlying service. For instance, instead of contenting himself with the vague ClustalW interface specification as described above, the domain modeler can simply specify two distinct alignment services, say `ClustalW_AA` and `ClustalW_NA`, of which one computes an amino acid alignment from a set of amino acid sequences, and the other a nucleic acid alignment from nucleic acid sequences. This *polymorphism* of services is in fact a very useful feature.

However, in the current implementation of the framework it often causes ambiguities with respect to the actual configuration of the synthesized workflows. The reason is that the difference between these services are only visible in the scope of the semantic domain model, but not at the level of the basic SIB instances. Thus, having a SIB in the workflow model, the framework can only guess which of the different service interfaces is to be applied.

Finally, the experience with the bioinformatics applications also showed that it is adequate to include only the actual input/output data that "flows" into/out of the service in the semantic service interface descriptions, but not the configuration parameters. While it is in principle possible also to regard the latter as inputs, this makes the interfaces and thus the synthesis tasks immediately more complex, which is undesirable with regard to trying to avoid state explosion problems.

7.2.3   Synthesis

Due to the impact of the characteristics of the domain model (see above), there is no ultimate synthesis strategy for all applications. Like no other framework known to the author, PROPHETS enables the user to "play" with configuration options and synthesis parameters and interactively explore the solution space. As discussed in greater detail in [233], PROPHETS provides many configuration options that can be used to tailor the applied synthesis strategy to the considered application domain. Further on, the work with the different bioinformatics application scenarios revealed some general rules-of-thumb for experimenting with synthesis parameters during the actual workflow design process, especially with search depth and constraints. In particular, it has turned out to be effective to:

- increase the search depth gradually instead of immediately searching for all solutions, and
- identify the required constraints incrementally instead of specifying everything from the beginning.

Initially, the synthesis is free to return all type-correct service compositions as solutions (from which then one is selected for the actual concretization of the loose model). This freedom needs to be limited by the user only when it is necessary, which is in particular the case when the obtained solution set is of unmanageable size. As such, workflow design in the scope of loose programming follows a notion of incremental formalization [303], and is hence in clear contrast to common computer science sense, where typically the goal is to provide a thorough and complete specification of the application before it is realized. This is, however, a demanding – if not impossible – task considering the variant-rich and varying objectives and framework conditions of the scientific workflow scenarios considered in the scope of this work that indeed rather ask for an ad-hoc workflow development style. Furthermore, the target users of the loose programming framework are typically neither able nor willing to provide thorough specifications in the classical computer science sense. Thus, the sketched approach to synthesis-supported workflow design is in fact most appropriate with regard to the envisaged user group and their applications.

Section 3.3.4 has already illustrated in detail the incremental constraint specification and search depth exploration for a particular workflow composition problem from the phylogenetics analysis workflow scenario. In the following, these aspects are revisited and regarded from a broader perspective in order to provide guidelines that are generally applicable.

Gradual Increase of the Search Depth

From a theoretical point of view, there are in the first place only two distinct sets of synthesis solutions that are interesting to consider:

1. the/one shortest solution,
2. all solutions.

Searching only for *one* shortest solution (typically simply the first that is encountered), however, is usually not sufficient in practice, as most applications comprise several adequate workflow variants, and not necessarily is a shortest solution the actually intended one. Thus it is reasonable to leave the selection of one particular solution to the user. However, finding *all* solutions is in practice usually not possible, as only very few synthesis universes can be explored completely (cf. Section 7.1.1).

Consequently, it is in practice reasonable to define an upper bound for the length of the considered solutions and start synthesizing with a small search depth, proceeding to greater depths only when it is clear that the solutions found in smaller ones are not sufficient. This strategy has also been applied in the detailedly described solution refinement strategy of the EMBOSS workflow scenario (cf. Section 3.3.4): In the unconstrained case, the default limit of 1,000,000 solutions (as defined by PROPHETS) has been reached while the synthesis algorithm was still exploring solutions of length 4. Taking this depth as the preliminary upper bound for the search, constraints were added until the synthesis algorithm only returned a manageable set of about 30 solutions. Only then, depth 5 of the synthesis universe was taken into account and explored for further admissible solutions.

As detailed in Section 7.1.2, the runtime performance of the synthesis can be roughly estimated as exponential, or $\mathcal{O}(m^d)$, where m is the number of services in the domain model and d the considered search depth. When trying to synthesize longer workflows based on more complex domain models, state explosion is likely to strike sooner or later and hamper the workflow design process. For this case, there is a simple but effective, divide-and-conquer-style workaround: perform two (or more) short syntheses instead of one long one, since $\mathcal{O}(m^{d1} + m^{d2}) \leq \mathcal{O}(m^{d1+d2})$. Typically, for longer workflows particular steps or services that are to be included can be identified beforehand, similar to looking out for suitable stones to step on when planning to cross a wild river. Instead of adding constraints that enforce their inclusion, these points can then be used to split a large synthesis problem into smaller, faster synthesizable ones.

Incremental Specification of Constraints

Instead of requiring to specify all characteristics of the intended application from the beginning, loose programming enables and encourages the user to identify the required domain- and problem-specific constraints incrementally. For this experimentation with constraints and obtained solutions, it has turned out to be advantageous to proceed in two principal phases:

1. define constraints that explicitly exclude inadequate solutions,
2. define constraints that explicitly include adequate solutions.

Often, the exclusion of undesired solutions already constrains the solution space clearly enough so that a manageable set of solutions remains. If this is not the case, providing more structure by explicitly describing parts of the intended solutions very effectively narrows the remaining solutions further. This strategy for the incremental specification of constraints has also been demonstrated for the EMBOSS workflow scenario (cf. Section 3.3.4): The first three of the finally six constraints exclude clearly useless (refinement step 1) and in the actual setting unwanted services (refinement step 2). The next three constraints explicitly include particular solutions by enforcing the use of specific services or service categories (refinement 3). In this particular example, a constraint set that specifies an assessable set of adequate solutions is reached at this point.

Note that especially the definition of constraints that include particular solutions comes with a certain risk of missing adequate solutions due to (unintended) over-constraining: It is easily possible to express requirements that can in their concrete form not be met by the domain model, although an acceptable solution to the actual synthesis problem principally exists. Hence, it is indeed advisable to leave as much freedom as possible to the synthesis algorithm, and to identify the adequate set of constraints incrementally.

Although the constraint identification process is only discussed explicitly for the EMBOSS workflow scenario, and not for the other three application scenarios (Chapters 4 – 6), it basically followed the same pattern there. Consequently, the domain constraints for these examples comprise both excluding and including constraints. The additional problem-specific constraints, which are provided by the workflow designer for a particular loose branch, are mostly constraints that include particular solutions, as at this point the user typically expresses particular intents about the solution that are too specific to be included in the general domain model.

8

Related Work

Due to massive effort in tailoring workflow management systems towards the bioinformatics domain and the development of thoroughly specific systems in the last years, several frameworks that support service orchestration in bioinformatics are available today. They provide a wide range of different features, adding convenience for the user in different ways. This chapter discusses the relation of Bio-jETI to other approaches to bioinformatics workflow management. Therefore, Section 8 reviews a number of different bioinformatics workflow systems and compares them to Bio-jETI. Then, Section 8.2 takes a closer look at the substantial differences between the capabilities of control-flow and data-flow modeling: The nature of the data flow and control flow specifications is essential for the semantic interpretation of a workflow model, while other features are usually less inherent in the systems. The study is carried out by means of a detailed comparison of Bio-jETI and Taverna, two systems for bioinformatics workflow management that look very similar at the first glance, but are in fact very different precisely because they follow different paradigms for control-flow and data-flow modeling.

8.1 Workflow Systems in Bioinformatics

Several approaches for meeting the service orchestration challenge in bioinformatics have been proposed and followed in recent years, ranging from simple scripting languages (like, e.g. Perl [17] or Ruby [13]) to sophisticated deductive systems (like, e.g. BioBike [282]). In particular, a number of dedicated workflow environments for bioinformatics applications has been developed, also following a wide range of different approaches and having different characteristics. This section gives an overview of more or less popular workflow management systems in the bioinformatics domain. Three groups of systems are considered: domain-independent workflow systems that have known applications in the bioinformatics domain (Section 8.1.1), workflow systems have been designed for the bioinformatics domain (Section 8.1.2), and workflow systems that have been made for a particular field of biological research (Section 8.1.3). Finally, Section 8.1.4 gives a survey of the presented systems and compares them to Bio-jETI with respect to the requirements for workflow management systems identified in Section 1.1.3.

Note that this selection of workflow systems raises no claim to completeness[1], but rather aims at giving a representative overview of the variety of systems that is available. Further reviews of workflow systems from different perspectives can be found, for instance, in [242, 106, 355].

8.1.1   *Domain-Independent Workflow Systems in Bioinformatics*

As a general review of domain-independent workflow systems would be beyond the scope of this work, this survey is restricted to domain-independent systems that have known applications in the bioinformatics domain. In contrast to many of the frameworks that are the subject of the following sections, these systems are usually based on sound software engineering techniques and thus they maintain the power of, for instance, model-driven design paradigms when they are used in the bioinformatics domain. These systems become domain-specific essentially by the definition of an appropriate domain model, that is, the integration of domain-specific services and the formal definition of semantic meta-information. Clearly, Bio-jETI also falls in this category of systems.

BPEL

BPEL (lately also termed BPEL4WS, Business Processes Execution Language for Web Services) [30, 140] is a framework for the composition of Web Services into processes, explicitly following the SOA (Serive-Oriented Architecture) paradigms. BPEL is the de-facto standard for business process modeling and has been implemented by a number of frameworks, for instance by ActiveVOS (formerly ActiveBPEL), Sun's OpenESB and IBM's WebSphere Process Server. Although being conceived for business process modeling, BPEL has also been used for the realization of scientific workflows (cf., e.g., [79, 117]).

BPEL workflows are composed from Web Services. The relationship between the services is defined by means of control-flow constructs (such as if-conditions, for-each- and while-loops) and activity containers (such as sequence and flow). The data is represented by XML documents, on which XPath and XQuery operations can be performed. An important notion of BPEL is to expose the entire workflow model as Web Service again, so that hierarchical composition is directly supported.

---

[1] A complete review is probably out of scope of any work. As of June 2012, Wikipedia's "bioinformatics workflow management systems" entry [340] alone lists 28 examples and includes references to further systems, and still covers only a small number of the actually available systems.

GNU Make

GNU make [8] is commonly known as a build utility that can be used for the automation of source code compilation processes. In the bioinformatics community it has also become a popular tool for producing pipelines and simple workflows using existing command-line tools as components [78]: make is able to handle and resolve dependencies between targets – hence, defining individual tool invocations behind the individual targets, and providing make with a final target will cause all required tools to be executed.

While GNU make saves programmers the work of defining workflows explicitly, makefiles are often confusing and in particular not accessible to people without programming experience. Domain-specific wrappers like BioMake/Skam [2] add some convenience for using GNU make for the definition and execution of bioinformatics workflows, but they do not increase the power of the approach: the automatic resolution of dependencies that is done by GNU make is at the end of the day only able to handle simple goal specifications and workflows of low complexity.

JOpera

JOpera [26], also sometimes called BioOpera [38], is an Eclipse-based platform that provides "process support for (more than) web services" [7]. It has been developed in the scope of a joint project of the Information and Communication Systems Research Group (IKS) and the Computational Biochemistry Research Group (CBRG) at ETH Zürich, Switzerland. The project aimed at improving and automating the large-scale analysis of genetic data sets [25].

JOpera applies the JOpera Visual Composition Language (JVCL) [245, Chapter 3] as a graphical notation for processes. Its essential components are service interfaces (representing, e.g., web services, Java methods, scripts, or subtasks) and their parameters. JOpera uses two models for specifying a workflow: the main process model defines the data-flow between the tasks and parameters, while a separate second model is used for more complex control-flow definitions between the tasks. The processes are compiled into (Java) executable code to achieve efficient execution.

SADI/SHARE

SADI (Semantic Annotated Discovery and Integration) [342, 343], primarily developed at the University of British Columbia in Vancouver (Canada), is a framework of conventions and best practices for the use of standard Semantic Web technologies for working with web services. Currently the main applications of SADI are in the life sciences, but the framework itself is conceived to be domain-independent. SHARE (Semantic Health and Research Environment) [331] is the current reference implementation of the SADI approach, comprising more than 300 annotated services and providing a simple web client for the specification and execution of workflows.

Conceptually, the key feature of SADI is to regard services as "annotators" that perform particular changes to the input data object and then return it as output. Technically, input and output of a SADI service are defined in RDF format in terms of OWL classes, and the behavioral semantics of the service simply corresponds to the transformation that is performed to get the output from the input OWL class. As a result of this, the OWL-DL domain models can be regarded as abstract workflow specifications [348], and standard OWL reasoners and SPARQL query engines can be applied to compose adequate services automatically. In particular, it is possible to describe the expected workflow outputs in terms of SPARQL queries, and then let the system determine and execute the (sequence of) services that transform the initial input to the final output.

Semantic Workflow Tool

The Semantic Bioinformatics Service Environment described in [256] is another approach that applies standard Semantic Web technologies for the description and composition of services: It uses OWL-S [214] for semantic descriptions of services, OWL for the definition of domain-ontologies, and a Description Logic (DL) reasoner for the automated discovery of services. In addition, the described environment comprises the Semantic Workflow Tool as a graphical user interface for (semi-) automatic workflow design and workflow execution. With the interface it is possible to construct service pipelines dynamically via semantic discovery of matching services, that is, by searching for a services that provide or use the input or output data of other services in the workflow.

The described environment focuses on Web Services as primary computational resource, and has been applied in the bioinformatics domain on popular publicly available Web Services. In order to provide a domain-specific framework, the originally domain-independent OWL-S ontology has been extended with ontological concepts from the bioinformatics domain, and then also used for the service annotations. In order to support interoperability and reusability, the Semantic Workflow Tool provides a functionality for exporting workflows in the Taverna workflow format.

8.1.2 *Workflow Systems for Bioinformatics*

Similarly, if not even more, interesting for comparison with Bio-jETI, however, are general-purpose workflow systems that have been designed for the bioinformatics domain [128]. In particular, the bioinformatics community has adopted the ideas of the Semantic Web earlier than other domains, and accordingly also systems for semantically guided workflow composition have been around for several years now. Without making a claim to be complete, the following gives an overview of the most popular approaches to workflow systems that have originated in the bioinformatics community,

BioSPICE Dashboard

The BioSPICE project [100], launched in 2001 as a collaborative initiative between the Lawrence Berkeley National Laboratory (LBNL), the University of California at Berkeley (UCB), the Stanford Research Institute International (SRI) and others, focused on the development of computational models for intracellular processes. Besides the provision of mere "access to the most current computational tools for biologists" [100], the work also aimed at integrating the services into a convenient graphical user environment, called the BioSPICE Dashboard.

BioSPICE builds upon the Open Agent Architecture (OAA) [67], a delegated computing architecture developed at SRI, for the integration of services and for the data transfer between the services during workflow execution. For the workflow definition, a NetBeans-based graphical user interface is provided. This interface also provides a simple but useful feature building upon basic semantic information about services, namely the categorization of services according to different criteria (in particular according to location, provider, function, or I/O type).

Discovery Net

The Discovery Net system [107], developed in the early 2000s in the scope of the UK e-Science Programme, is one of the earliest workflow systems that have been designed specifically for scientific purposes. As several of its features (such as the graphical interface and access to remote Web and Grid services) were considered novel at that time, it served as a role model for the development of other scientific workflow systems in the following years [106]. The system has been used in a number of bioinformatics projects, especially in the UK (cf., e.g., [108, 76]).

Discovery Net has been designed for workflow management in a grid computing environment, and thus provides the means for integrating and orchestrating distributed resources and for deploying workflows for execution on grid infrastructures. As for its functional range, Discovery Net can be classified as a closed-world system [106], since it provides a dedicated set of components for data mining tasks, but is not prepared for integrating (arbitrary) new services. In [75] an approach for the formal verification of Discovery Net models based on temporal logic is proposed. However, these analyses are carried out on a state model that is extracted from the actual workflow model, rather than directly within the workflow environment, as it is the case in Bio-jETI (cf. Section 2.2).

Galaxy

Galaxy [114, 45] is a web-based workflow system targeted at data-intensive bioinformatics analyses. It has been developed jointly by the Center of Comparative Genomics and Bioinformatics at Pennsylvania State University and the Departments of Biology and Mathematics and Computer Science at Emory University. Recently it has predominantly been used for genomic research (cf., e.g., [156, 46, 44]).

The Galaxy web interface provides a rich (and extensible) collection of computational tools for various general and bioinformatics tasks and predefined access methods to common bioinformatics databases. Upon workflow execution, Galaxy automatically creates and stores provenance information and metadata about the executed tools and workflows. As an open web platform, it furthermore functions also as community space for sharing workflows, data, and results, and provides the possibility to equip tools and workflows with tags and annotations.

jORCA

jORCA [215, 147], developed at the University of Malaga (Spain), addresses interoperability issues of bioinformatics web services that arise from the use of incompatible data and communication formats. Therefore it combines the integration of heterogeneous services into one uniform framework with service discovery and automatic workflow composition functionality. jORCA integrates web services from a number public repositories, such as the BioMoby [341], DDBJ [165] and EBI [250, 167, 118] web services. As usability is a major goal of jORCA, its user interface is conceived as desktop client that allows users to "compose, edit, store, export, and enact workflows" [147].

The automatic workflow composition functionality is based on the Magallanes system [265], which uses a breadth-first pruning algorithm (cf. [85]) for finding a shortest path from source (i.e., input data type) to target (i.e., output data type). Thus, jORCA can create input-output pipelines automatically, but in contrast to the temporal-logic synthesis methods applied by PROPHETS, it is neither able to handle additional constraints, nor to use a shared memory for more complex data transfer.

Kepler

Kepler [28] has been designed as a system for speeding scientific workflows in the scope of a cross-institution project at UC Davis, UC Santa Barbara, and UC San Diego. It has been applied to different domains, such as astronomy, ecology, geology and biology (cf., e.g., [53, 126]).

Extending the Ptolemy II system [91] for heterogeneous, concurrent modeling and design, one of Kepler's most notable characteristics is its support for different models of computation for the workflow models. For instance, synchronous data-flow enactment can be chosen as the adequate model of computation for simple data transformation workflows, whereas continuous-time enactment is suitable for time-dependent workflows where the model is described in terms of differential equations. Workflow development in Kepler takes place in a graphical user interface that provides a number of preconfigured local tasks and configurable Web service clients. What is more, the Kepler Analytical Component Repository [11] maintains a collection of analytical services that can be downloaded and imported into Kepler.

Taverna

Taverna [134] is a workflow environment for the life sciences that has been developed in the scope of myGrid [309], a UK initiative focusing on eScience. Not least because it is strongly supported by the UK life science community, it is currently the probably most popular workflow system in bioinformatics. Consequently, Taverna has been used within a large number of projects, predominantly in bioinformatics (cf., e.g., [184, 185, 195]).

Taverna provides a simple, graphical user interface for building and executing workflows. In addition to a small set of locally operating workflow building blocks for common tasks, Taverna integrates bioinformatics web services and other kinds of remotely available tools, explicitly following the open-world service assumption (cf. Section 1.1.3). Most notably, Taverna includes comprehensive mechanisms for the automatic discovery of web services for bioinformatics (browsing, e.g., the BioMoby registries and EBI's SoapLab), which makes it very convenient to use for workflow designers that are not willing to or not capable of integrating the services themselves.

At this point also the myExperiment project [113] of the myGrid initiative should be mentioned, which aims at bringing the Web 2.0 spirit into the life science community. The platform enables researchers to publish, share, find, and download workflows, with the aim of making the re-use of existing workflows as easy as possible. They can furthermore tag, rate, discuss and comment *in silico* experiments, create and join special interest groups, and connect with people. At present, most workflows that are available in myExperiment are in fact Taverna workflows (cf., e.g., [189]), but the platform is in principle open to any other workflow format.

Taverna with SADI Plugin

The SADI plugin [344] for Taverna (cf. Section 8.1.2) extends the workbench by functionality for interactive service discovery: It makes it possible to search the SADI registry (see above) for services that consume the data type that is the output of an existing workflow node. However, although the SADI plugin provides means for semantically guided discovery of matching services within Taverna, it does not bring the full power of the SADI system into the workbench.

Note that the UK eScience initiative myGrid [309], where Taverna originated, has also investigated the use of Semantic Web standards for their projects [190] and for instance developed the myGrid Ontology for facilitating bioinformatics service discovery [347] . However, to date they have not visibly been applied for semantics-based, automatic workflow composition.

Triana

Triana [318] is a workflow environment that has been developed at Cardiff University, UK. It has in particular been designed to support building grid workflows in physics and astronomy, but is in principle domain-independent and has also been used to implement bioinformatics workflows (cf., e.g., [217]).

Triana provides a comprehensive set of ready-to-use, powerful data analysis tools for tasks like signal analysis and image manipulation, and is extensible to other, for instance, Web, grid, and P2P services, which reflects Triana's aim of integration with grid technologies. The services are connected graphically in order to define the data flow, but dedicated nodes for common control-flow structures are provided in order to achieve basic support for additional control.

Wildfire

Wildfire [315] has been developed by A*STAR Bioinformatics Institute, Singapore. It serves as a graphical interface for defining GEL workflows. GEL (Grid Execution Language) [186] is a scripting language for describing parallel workflows, which can also be executed on Grids or cluster machines. Primarily, Wildfire provides interfaces to the tools of the EMBOSS suite, but it is possible to extend the component library to comprising further command-line programs. Workflows are simply defined by connecting components with pipelines that define the flow of data.

Wings

Pegasus is "a framework for mapping complex scientific workflows onto distributed systems" [83] that has been used in a number of applications from different scientific disciplines, also including bioinformatics. The Wings (Workflow INstance Generation and Selection) system [111] that has been fully integrated into Pegasus provides functionality for (semi-) automatic workflow creation based on semantic representations and planning techniques. Wings puts a strong focus on the analyzed data, that is, intermediate results can influence the generation of subsequent workflow parts, and semantic descriptions of results are automatically generated. The system distinguishes three distinct stages in workflow development, namely the creation of workflow templates (where an expert user specifies high-level, data-independent workflow templates), the creation of workflow instances (where the normal user specifies the data that is to be used by the computation) and the creation of the actually executable workflows (where the workflow instance is automatically mapped to a concrete execution environment by Pegasus).

Indeed, Wings supports loose specification, concretization, and also constraint-based validation of workflows in a way that reminds of the loose programming approach followed by Bio-jETI/PROPHETS. However, within Wings the loosely specified workflow templates as well as all workflow constraints are already defined as part of the domain model by an expert user, while the normal user can only build workflow instances by specifying the data that is to be used. This is in contrast to Bio-jETI, where the average user is regarded as workflow designer who works with a comprehensive, prepared domain model, but can create completely new workflows on his own and can also refine the domain model by adding his intents or further expert knowledge.

8.1.3   Workflow Systems for Particular Bioinformatics Applications

Another kind of bioinformatics workflow systems are tools that provide workflow support for very special purposes, such as particular biological disciplines or the bioinformatics applications that are supported by a particular platform. Technically, they follow a variety of approaches highly similar to the general systems presented above. In case that automatic service composition functionality is provided, it is naturally limited to the semantically annotated services of the respective platforms.

BioMoby

BioMoby [341], initiated in 2001 at the Model Organism Bring Your own Database Interface Conference (MOBY-DIC), is a service registry that has been developed specifically for the bioinformatics domain. The actual registry, MOBY Central, holds the services and their interface descriptions. In addition, BioMoby comprises service and object hierarchies that describe the relationship between the involved service and data types, respectively. These simple ontologies provide a BioMoby-specific vocabulary that can be used by service providers to annotate their services when registering them at MOBY Central.

Several clients support semantically guided service composition based on BioMoby. With the MOBY-S Web Service Browser [84] it is, for instance, possible to search for an appropriate next service and store the sequence of executed tools as Taverna workflow, the REMORA web server [64] offers functionality for the discovery and step-by-step composition of matching BioMoby services, and the SeaHawk browser [116] facilitates the data-centric composition workflows that are then executed in Taverna. However, since development and maintenance of BioMoby have been abandoned in favor of the standards-compliant SADI framework (see above), their future use is questionable.

DDBJ Workflow Navigation System

The DDBJ's Workflow Navigation System [165] is a web-based platform that works on the DDBJ [165] and NCBI e-Utils [270] web services. It allows users to perform workflows on the fly, that is, applying multiple services consecutively within the web browser, whereby the system assists in finding applicable subsequent services based on the outputs of the already executed services.

GenePattern

GenePattern [263] is a web-based platform maintained at the BROAD Institute, which focuses on the analysis of genomic data via reproducible in silico experiments. Therefore, it provides access to more than 125 specialized tools and general data processing tasks and supports the definition and execution of simple analysis workflows (pipelines).

Proteomatic

Proteomatic [291] is a recently released framework specializing in workflows for tandem mass spectra (MS/MS) data analysis, which has been developed at the Institute of Plant Biology and Biotechnology, University of Münster. Workflow building blocks in Proteomatic are simply scripts that access locally installed tools for particular proteomics analyses. Workflows can be constructed and executed from the command line or via a graphical user interface.

RSEQtools

RSEQtools [123] is a software suite targeted at the analysis of data from RNA sequencing (transcriptome profiling) experiments. that has been developed in the scope of a joint project of computer science, bioinformatics, molecular biology and medical research groups at Yale, Harvard and Stanford. The tools in the suite are provided in a form that allows for the easy invocation via a command line interface and thus also enables the creation of simple, customizable workflows using standard scripting languages.

SeaLife Argumentation Interface

The *SeaLife* project [277] aims at establishing a Semantic Web browser for the life sciences (currently especially for the study of infectious diseases) based on Semantic Web technologies and the available eScience infrastructure of the field. One example of the context-based information integration that is envisaged by the project is the *SeaLife Argumentation Interface* [313], where meaningful terms from the gene expression domain are recognized in the text of a web page and used for the formulation of higher-level goals, which are, together with web services that are linked to the terms, given to an HTN (Hierarchical Task Network) planner [231, 284] that then creates workflows that are suitable within the current context.

8.1.4   Survey and Comparison with Bio-jETI

Although only covering a small selection of systems, this overview gives a representative impression of the landscape of workflow management technology that is available for bioinformatics application. In particular, it makes clear that the systems differ in their focus on specific aspects of workflow design (often because of originating from a particular field of research), and that no single system fits all requirements completely. Interestingly, several systems focus on semantics-based functionality, but do not address the basic requirements for workflow systems adequately, and vice versa. Naturally, this limits their applicability considerably.

Bio-jETI addresses the basic requirements for workflow systems as well as the demands for semantics-based enhancements. Furthermore, it differentiates itself from other bioinformatics workflow systems in particular through its focus on control-flow modeling and through its workflow validation and verification facilities.

Table 8.1 surveys the systems presented above systematically with respect to the requirements listed in Section 1.1.3. The + indicates that the respective requirement is met by the system in a satisfying way, the – denotes that it is not sufficiently realized, and the o is used when the requirement is only partially fulfilled. A ++ is used when the respective system puts particular emphasis on this aspect, which typically means that the corresponding requirement is addressed particularly well. Without going into the details of the single properties, the table backs the following observations about the considered systems and their relation to Bio-jETI with regard to the requirements.

Abstraction (Requirement 1)

Abstraction from programming details is central to most systems. That is, the systems clearly separate the workflow definition from its execution, and are thus able to abstract from the underlying programming language code and technical details, so that the user can deliberately focus on the service orchestration. Not surprisingly, those systems that furthermore provide comprehensive but nevertheless intuitive desktop or web-based graphical user interfaces have turned out to be most successful.

In Bio-jETI, abstraction is achieved by the rigorously service-oriented conception of the SIBs and SLGs, and by the intuitive graphical user interface of the jABC.

Powerful Workflow Model (Requirement 2)

While most bioinformatics systems focus on the data (flow) handling in the workflow models, the domain-independent systems usually consider a focus on control-flow modeling more adequate. In Bio-jETI, control-flow handling

Table 8.1 Survey of workflow systems in bioinformatics

System	R1: abstraction	R2.1: control-flow handling	R2.2: data handling	R2.3: hierarchy	R3.1: service integration	R3.2: semantic domain model	R4.1: service discovery	R4.2: automatic workflow composition	R5.1: validation/verification of static properties	R5.2: validation/verification of runtime behavior	R6.1: execution (internal)	R6.2: execution (external)
Bio-jETI	+	+	o	+	+	++	+	++	+	+	+	+
BPEL	+	+	o	+	o	-	o	-	o	o	o	o
GNU make	-	+	o	+	+	o	o	o	-	-	+	-
JOpera	+	+	+	+	+	-	-	-	o	o	+	o
SADI/SHARE	+	-	+	-	o	+	+	++	-	-	+	-
Semantic Wf. Tool	+	-	+	-	o	+	+	-	o	o	+	-
BioSPICE	+	-	+	o	o	o	-	-	o	-	+	o
Discovery Net	+	+	+	+	+	-	-	-	o	o	+	-
Galaxy	+	o	+	o	o	o	o	-	o	o	+	-
jORCA	+	-	+	-	+	+	+	+	o	o	+	-
Kepler	+	o	+	+	+	-	-	-	o	o	+	o
Taverna	+	o	+	+	++	-	o	-	o	o	+	-
Taverna + SADI	+	o	+	+	++	o	+	+	o	o	+	-
Triana	+	o	+	+	o	-	-	-	o	o	+	o
Wildfire	+	-	+	+	o	-	-	-	o	o	+	-
Wings	+	-	+	+	+	+	+	+	o	o	o	+
BioMoby	+	-	+	-	-	+	+	o	o	o	+	o
DDBJ Wf. Navi.	o	-	+	-	-	o	+	-	-	o	+	o
GenePattern	o	-	+	-	-	-	o	-	o	o	+	-
Proteomatic	+	-	+	o	o	-	-	-	o	o	+	-
RSEQtools	-	o	+	+	o	-	-	-	-	o	+	-
SeaLife	+	-	+	-	-	+	+	+	o	o	+	-

is inherent in the SLGs, which are control-flow representations of workflows, while data handling is done via the execution contexts. In fact, it is part of the jABC philosophy to hide the data (flow) to a certain extent behind the behavioral workflow model: the control-flow is represented by the graphical workflow model, while the data-flow is defined via component configurations. Thus, data dependencies between the single services do not obstruct the workflow representation, and even large workflow with complex data flows are still easily readable. In contrast, the data-flow approach that is taken by the majority of popular bioinformatics workflow systems is based on the philosophy that the data is central to the analysis workflows and have thus to be put in the foreground of the workflow definition. However, as Section 8.2 will discuss in detail, control-flow-oriented workflow models like Bio-jETI's support the definition of workflows with greater computational complexity. At the end of the day it depends both on the concrete application as well as on the preferences of the workflow modeler which data handling concept is more appropriate.

Most systems support hierarchical workflow models that allow for encapsulation of sub-workflows into reusable components. In Bio-jETI, complex workflows are enabled by the hierarchy concept of the SLGs.

Domain Modeling (Requirement 3)

The vast majority of systems share the "open world service assumption" [242] and allow for the extension of the component libraries. However, support for (semi-) automatic integration and service management is realized in very different ways.

In Bio-jETI, the open world assumption for services is met by the technical concepts of the SIBs (freely extensible Java classes), and additionally supported by the jETI platform and the SIBCreator plugin. Service management via the SIB browser is supported by the Taxonomy Editor plugin, while additionally the PROPHETS plugin provides the means for semantic service management and transparent discovery.

Most systems provide no means for the semantic enrichment of the domain model at all. A couple of systems support basic semantic domain modeling, but rely on predefined domain models that are only supposed to be changed by their developers, that is, they use the knowledge that is provided by the service and data type descriptions and ontological classifications, whereas all additional domain knowledge (if any) is hidden in specifically designed composition algorithms. Wings defines the creation of the workflow templates as distinct phase, which is part of the domain modeling together with the component and constraint specifications. GNU make, SADI and the Semantic Workflow Tool can deal with changing domain models, but they do not provide explicit support for domain modeling.

Bio-jETI is the only system among the presented that offers comprehensive semantic domain modeling support (including the both definition of domain ontologies and service interface descriptions) and explicitly gives the user the possibility to adapt predefined domain models according to his expert knowledge or specific requirements. What is more, it enables a very flexible way of expressing additional knowledge that is cleanly separated from the implementation of the synthesis algorithm.

Semantics-Based Service Composition Support (Requirement 4)

If service composition support is provided, in most systems it is restricted to single steps of the workflow, that is, service discovery: the systems support the identification of services that can provide or use the data that are used or provided by services that have already been inserted into the workflow. This semantically supported step-by-step workflow construction is actually rather service discovery than real automatic workflow composition, and comes with the risk that users get stuck when trying to construct the globally intended solution stepwisely.

Truly comprehensive abstract workflow descriptions are only supported by Bio-jETI (SLTL specifications), SADI/SHARE (arbitrarily complex SPARQL queries) and the SeaLife Argumentation Interface (higher-level planning goals). The final target specification used for automatic workflow creation with GNU make, the start-to-end specification used by jORCA and the workflow templates that are applied within Wings are less expressive, but can still be regarded as abstract workflow descriptions.

Workflow Validation and Verification (Requirement 5)

Verification and validation of workflow models is mostly realized on a purely syntactic level (e.g. syntax or type checking on the component level), whereas semantics-aware verification methods that consider the entire workflow model (e.g. model checking) are provided only by very few systems. In Bio-jETI, validation and verification methodology is available via the LocalChecker and the GEAR model checking plugins, respectively. They enable constraint-driven workflow design by continuous monitoring of the workflow development process in terms of the constraints that are defined within the domain model. To the best of the author's knowledge, no other workflow system in bioinformatics provides similarly integrated verification and validation facilities.

Regarding the workflow systems considered here, the application of model checking techniques has only been described for the analysis of Discovery Net workflows [75]. However, there the model checking is not performed directly on the workflow model within the workflow modeling framework, but as a separate step on a specific, derived representation. This means in particular that the results of the analysis can not be adequately used to support the workflow design phase, as they only provide information about the (finally) exported workflow model. Moreover, the examples provided in [75] merely show how general properties of the workflow models (such as reachability of certain services and potential deadlocks or livelocks) are detected. Due to incorporating aspects that are more specific for the application domain and the specific workflows, the constraints used within Bio-jETI provide useful information for the workflow design process, other than the generic properties described in [75].

Means for runtime validation, a.k.a. debugging functionality, are offered by most systems via the workflow interpretation facilities (see below) that provide stack traces or other error messages when execution errors occur. In Bio-jETI, workflow debugging functionality is provided by the Tracer plugin.

Workflow Execution (Requirement 6)

At least basic workflow execution facilities are provided by all considered systems. Several systems also support the deployment of the workflow model into specific (grid) execution language. More general model compilation functionality is basically only available in the domain-independent systems.

In Bio-jETI, workflow execution by interpretation as well as workflow debugging functionality is provided by the Tracer plugin. Comprehensive workflow compilation and deployment support is provided by the Genesys code generation framework.

8.2 Control-Flow or Data-Flow Modeling?

The most substantial difference between workflow management systems is typically the nature of their process models, that is, whether they express the data flow or the control flow between the building blocks. In data-flow modeling, the connections between building blocks are interpreted as data pipelines, and execution control follows the data flow implicitly. In control-flow modeling, the connections between building blocks define the flow of control, and thus the data flow is modeled separately. While the nature of the data-flow and control-flow specifications are essential for the semantic interpretation of a workflow model, other features are usually less inherent in the systems, and are thus more often subject to changes during further development of the software.

The case study that is presented in this section (and which has in parts also been discussed in the scope an encyclopedia article on bioinformatics workflows [176]) aimed at examining the differences between data-flow- and control-flow-based systems especially in the context of bioinformatics workflows. Therefore three increasingly complex reference workflows (details in Section 8.2.1) have been selected that are suitable as benchmarks for the modeling capabilities of the different software systems.

Table 8.2 surveys the workflow systems discussed in Section 8.1 with respect to the nature of the workflow models, that is, whether they focus on data-flow modeling, on control-flow modeling, or whether they apply an explicit hybrid approach that combines both kinds of models. It reflects the already mentioned tendency of specifically developed bioinformatics workflow systems towards data-centric modeling. Note, however, that most systems do not base their workflow models on formal definitions, and that often data-flow models are extended by particular control-flow features, and vice versa. Hence one cannot assume per se that general results regarding the capabilities of control-flow and data-flow structures (e.g., [321, 152, 262]) apply.

The study was carried out by realizing the selected benchmark workflows in Bio-jETI (Section 8.2.2) and Taverna (Section 8.2.3), which are characteristic representatives of control-flow- and data-flow-based systems, respectively. Both approaches are often considered to be capable of expressing the same workflows, but in fact there are limitations with respect to the inclusion of elaborate control structures when using the data-flow approach. Consequently, results show that linear and parallel workflows can be realized by both frameworks without problems. In Taverna, the realization of conditional branching is already difficult, while workflows that involve dedicated iterations or parameter sweepings can sometimes not be realized at all. The control-flow based models of Bio-jETI, on the other hand, also cover these more complex workflow structures.

The study was carried on by Sariette Bille in her diploma thesis [42], where she systematically analyzed a selection of DDBJ workflows [4] with respect to the general workflow patterns described in [326]. In essence, her results confirmed the findings of the initial study. Analogous results have also been obtained by similar studies (e.g., [52, 281, 106, 336, 314, 117]), all together covering a wide range of scientific and business workflow systems.

Table 8.2 Data-flow, control-flow, and hybrid systems

control-flow systems	data-flow systems	hybrid systems
BPEL Bio-jETI GNU make RSEQTools	BioMoby clients BioSPICE Dashboard DDBJ Workflow Nav. System Galaxy GenePattern jORCA Kepler Proteomatic SADI/SHARE SeaLife Argumentation Interface Semantic Workflow Tool Taverna Triana Wildfire Wings	JOpera DiscoveryNet

8.2.1   Reference Workflows

For the comparison of Taverna and Bio-jETI with respect to their workflow modeling capabilities three increasingly complex reference workflows have been selected. They are introduced in the following, before their realizations in both workflow systems are discussed. All examples are third-party workflows in order to make sure that the characteristics of the benchmarks have not been influenced by the system under consideration beforehand. In particular, examples from the myExperiment workflow repository [113], for instance, which primarily contains Taverna workflows, would not have been suitable for this purpose.

Workflow Example 1: Multiple Sequences Analysis

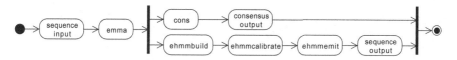

Fig. 8.1 Multiple sequences analysis workflow

The first example workflow (sketched in Figure 8.1) originates from the WebLab project [349]. WebLab is a web-based platform for executing and combining bioinformatics analysis tools, predominantly of the EMBOSS suite (cf. Section 3.3). Therefore, this example is mainly based on EMBOSS tools. The workflow is basically linear, but as the figure shows can easily be parallelized, as intermediate results are used independently by different subsequent services.

Taking a set of (nucleic or amino acid) sequences as input, the workflow computes their consensus sequence and generates set of corresponding new sequences. Therefor it starts with a multiple sequence alignment computed by an EMBOSS program called **emma**, which is an interface to the ClustalW algorithm. A consensus sequence is calculated from the alignment using **cons**. The alignment is also used as input for **ehmmbuild**, which builds a profile Hidden Markov Model (HMM) from the alignment. The HMM search statistics are calibrated by **ehmmcalibrate**, before **ehmmemit** is used to generate a set of sequences from the HMM.

Workflow Example 2: Phylogenetic Tree Construction

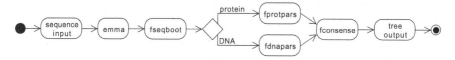

Fig. 8.2 Phylogenetic tree construction workflow

Like the first example, also the second workflow (see Figure 8.2) originates from the WebLab project and is also based on EMBOSS tools. In addition to the linear and parallel execution of service sequences, realizing this workflow required conditional branching at a point where the appropriate tool has to be chosen according to properties of the workflow input.

The workflow takes a set of (nucleic or amino acid) sequences as input and finally procuces a phylogenetic tree. The first step computes a multiple alignment of the input sequences using emma. Then fseqboot is run on the alignment in order to generate multiple data sets that are resampled versions of the input data (this is useful later to evaluate the significance of the consensus tree). Depending on the type of the input sequences either fdnapars (for nucleic acid sequences) or fprotpars (amino acid) is applied next, estimating phylogenies by the parsimony method. Finally, a strict consensus tree is derived from the phylogenies using fconsense.

Workflow Example 3: DDBJ-UniProt Workflow

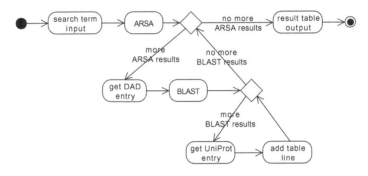

Fig. 8.3 DDBJ-UniProt workflow

The third example (sketched in Figure 8.3) that was applied in this study has been listed as a sample workflow during the 2008 BioHackathon [148], which focused on bioinformatics web services and workflows. Realizing this process needs a more sophisticated control-flow handling than the previous examples, since it requires the use of at least two nested for-each loops.

The workflow input is a search term (for example a protein name), the final output a table that contains aggregating DDBJ and UniProt information corresponding to the search term. Therefor, the input search term is used for a query to the ARSA seach service, which returns a list of accession numbers that match the query. For each accession number, the corresponding sequence is fetched from the DDBJ's DAD database (in FASTA format) and "blasted" against the UniProt database. For each blast hit, the corresponding UniProt entry is retrieved in order to extract the Protein ID and description. This information is used to successively fill a table with accession numbers, protein IDs, UniProt IDs, and protein descriptions. Thereby data from all scopes (not just from the inner loop) is required for assembling the table data.

8.2.2   Control-Flow Realization

The Bio-jETI framework, as described in Chapter 2, was used in this study for control-flow realizations of the benchmark workflows. The Bio-jETI version that was used is based on jABC version 3.7 and jETI 1.3 (as released in October 2008). As detailed before, the jABC is clearly a control-flow-oriented environment, providing a set of different common control structure in the form of SIBs, and expressing the control-flow via the connection between the single SIBs, while the application data itself is managed within the associated execution contexts.

Example 1 in Bio-jETI

The building blocks that are needed for assembling the multiple sequences analysis in Bio-jETI are, on the one hand, SIBs that act as clients to the involved EMBOSS services, and on the other hand SIBs that perform local tasks like input handling and result representation. While a comprehensive set of SIBs for common local tasks is provided by the jABC, the specific web service clients have to be created first.

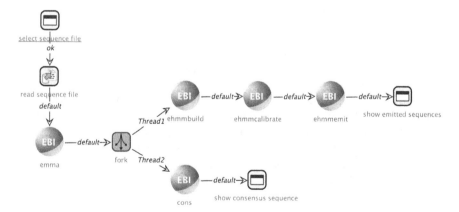

Fig. 8.4 Multiple sequences analysis workflow in Bio-jETI

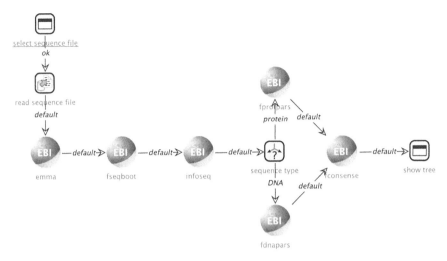

Fig. 8.5 Phylogenetic tree construction workflow in Bio-jETI

Figure 8.4 shows one possible realization of the multiple sequence analysis process in Bio-jETI: The input is read from a file that is specified by the user at runtime. After the alignment has been computed by `emma`, the control flow splits into two parallel threads. The first thread calls `ehmmbuild` and then splits again into two subthreads for calling `ehmmcalibrate` and `ehmmemit`, respectively. The second thread calls `cons` and displays the resulting consensus sequence. Note that it is not necessary to run these tasks in parallel, sequential execution of the same tasks is also possible and would lead to the same results.

Example 2 in Bio-jETI

The process for the phylogenetic analysis (Figure 8.5) makes use of similar SIBs. It begins by asking for a file name and reading the input sequences from that file. After applying `emma` and `fseqboot` to the sequences and the alignment, respectively, the workflow splits depending on the type of the input sequences. For the user's convenience this part of the process was realized in such a way that the input classification is not part of the input, but is rather automatically derived from the sequences themselves. Therefore, the `infoseq` service is called, which returns, among other information, the type of the input sequences. The `infoseq` result is then evaluated by a SIB which directs the flow of control to `fdnapars` in case of a nucleic acid sequence and to `fprotpars` in case of a amino acid sequence. The subsequent process is then again the same for DNA and protein, so both branches direct the flow of control to `fconsense` and finally to a SIB displaying the result.

Example 3 in Bio-jETI

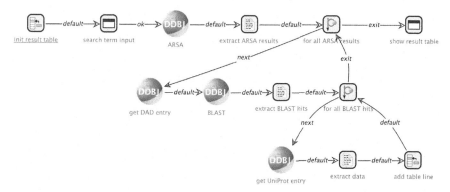

Fig. 8.6 DDBJ-Uniprot workflow in Bio-jETI

The third example is based on the DDBJ services, which have already been used in the examples of Section 3.2. At the beginning of this workflow (top left), an empty table is initialized, where during the workflow run the results are collected. Then, the ARSA search service is called and the database IDs of the matches are extracted from the result. The iteration over these IDs constitutes the first loop. Within this loop, the entry corresponding to the current ID is fetched and "blasted" against a database, again resulting in a list of entries (represented by their IDs). Iterating over these IDs yields a loop within a loop. Here, the database entries corresponding to the BLAST hits are successively retrieved, and their IDs and descriptions are added to the result. The resulting table finally consists of the IDs returned by ARSA (first column), the IDs found by the BLAST search (second column) and the corresponding ID and description fields of the actual database entries (remaining columns).

8.2.3 Data-Flow Realization

For the realization of the benchmark workflows in a data-flow based environment, Taverna version 2.2.0 (released in July 2010) was used. Very similar Bio-jETI, Taverna supports graphical workflow modeling on a central canvas, where services are orchestrated into workflows. Figure 8.7 gives an impression of Taverna's graphical user interface for the workflow construction (the so-called Design Perspective): The design window consists of three basic parts: a list of the available local and remote services (top left), a table managing the services and *links* that constitute the workflow model (bottom left), and a graphical representation of the workflow on the right. In contrast to Bio-jETI, the branches between the (input and output ports of the) single services define a data-flow model of the process. In order to influence the flow of control

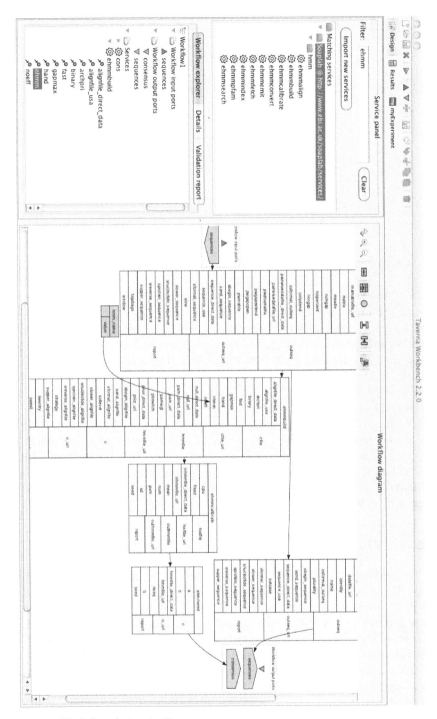

Fig. 8.7 Workflow design in Taverna

further, simple dependencies can be expressed by defining additional *control links*, or by defining iteration strategies for single components.

Taverna workflows can directly be executed once all involved services are connected. The workflow execution is managed in a separate window, in the so-called Results Perspective, which takes care of workflow execution, documents the progress and status of its execution in detail, and presents the (final and intermediate) results. Upon execution, each of the specified workflow inputs has to be provided with data, the outputs are then listed in the results perspective, from where they can be processed further.

Example 1 in Taverna

Coming to the realization of the workflow for analysis of multiple sequences in Taverna, the first step is to search the list of available processors for the appropriate EMBOSS tools. As Taverna's discovery mechanism browses the services provided by EBI's SoapLab [14, 15] automatically, they are directly available and can be added to the model. Furthermore a a workflow input component is needed that represents the multiple sequences that the workflow deals with, and three output representatives for the consensus sequence, the HMM and the emitted sequences.

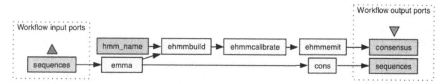

Fig. 8.8 Multiple sequences analysis workflow in Taverna

The data links are defined as can be seen in Figure 8.8: The initial sequences are sent to **emma**, the resulting alignment is the input for both **cons** and **ehmmbuild**, and the constructed HMM is sent to **ehmmcalibrate** and **ehmmemit**. For **ehmmbuild** a name for the constructed HMM must be specified, which is done via the String constant processor **hmm_name**. The finally resulting consensus sequences and HMM-emitted sequences are directed to the workflow output components **consensus** and **sequences**, respectively.

Example 2 in Taverna

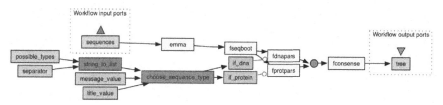

Fig. 8.9 Phylogenetic tree construction workflow in Taverna

As the tools required for the second example are also part of the EMBOSS suite, they are available as processors in the same way. Similar to the first example the workflow for constructing a phylogenetic tree (see Figure 8.9) starts with sending a set of sequences to `emma`. The resulting alignment is then forwarded to `fseqboot`, which generates a set of reference sequences. Now the continuation of the analysis depends on the type of the input data: nucleic acid sequences have to be processed by `fdnapars`, amino acid sequences by `fprotpars`.

The conditional branching that is required here can be realized in Taverna by using control links and an additional (boolean) input specifying if the input is DNA or not. This boolean value is (data-) linked to two processors, `fail_if_true` and `fail_if_false`, which are (control-) linked (the gray connectors with the circle in Figure 8.9) to `fprotpars` and `fdnapars`, respectively. Only one of `fail_if_true` and `fail_if_false` can be successfully passed during execution, and subsequently either `fprotpars` or `fdnapars` has both input links available and will be executed. The remainder of the workflow is straightforward: the resulting phylogeny is directly used as output and additionally sent to `fconsense` in order to compute the consensus tree from the large number of possible results.

Example 3 in Taverna

Although the DDBJ web services on which this example is based can be accessed via Taverna processors, it is not possible to realize the whole process in the workbench. The reason is that the aggregation of information for the resulting table takes place in the inner of two nested for-style loops: The outer loop iterates over the list of accession numbers that are returned by ARSA, and the inner loop processes the results of the BLAST search that is performed for each accession number.

While this so-called parameter sweeping can be built in control-flow-based systems in a straightforward manner, it has not been possible to implement it in Taverna's data-flow-oriented environment when using only the functionality that is readily provided by the workbench. The implicit loops that are usually provided by the latter are sufficient for batch-like processing of datasets, but fail when the well-directed addressing of data items within the loop is necessary.

8.2.4 Summary

Taverna's focus on easy, automatic import of common bioinformatics services, in combination with its data-flow-based workflow models, makes it convenient for rapid development of batch processing jobs and computationally simple workflows. Modeling the data-flow, the workflow developer takes a resource-oriented perspective, which is close at the actual data that is processed by the workflow. More complex control-flow constructs, like conditional branching or looping, can however be difficult or impossible to realize.

In contrast, Bio-jETI emphasizes the computational power of its workflow models and provides thorough control-flow structures, which are necessary for more complex analysis problems. The control-flow-oriented approach furthermore allows workflow developers to think in "Dos" and "Don'ts" and steps and sequences of action in their own terms at their level of domain knowledge. The data-flow is modeled separately in Bio-jETI following a shared-memory approach. In case of computationally simple workflows, this leads to the problem that some users feel they were doing the same work twice, namely when they draw branches between SIBs and have to configure their parameters so that the data is passed from the one SIB to its successor. Readily configured linear (sub-) workflows can, however, be automatically generated by using the synthesis feature of the PROPHETS plugin given that an appropriate domain model is available.

Naturally, the choice for a particular workflow system always depends on the concrete use case and to some extent also on the skills and preferences of the workflow developer. As discussed above, however, the availability of control flow structures and flexible data exchange facilities (such as a shared memory) is necessary when it comes to analysis problems with certain computational complexity. In fact, it has been pointed out already in [336] that most "workflow systems in this domain lack the facilities to model advanced control structures such as conditional branches and iteration", and that "there is growing demand for a more controlled approach to workflows in the life science domain". Hence, systems like Bio-jETI, which make use of control-flow workflow models, are generally preferable.

9

Conclusion

This book has addressed the development of a framework for user-level work-flow design, its application to different bioinformatics application scenarios, and the evaluation of the approach in itself and in comparison to related work. This chapter concludes by summarizing the work and clarifying achievements and major remaining challenges in Section 9.1, before outlining directives for future work in Section 9.2.

9.1 Summary

The general question addressed by this book can be phrased as:

How can user-level design of workflows be achieved?

One possible answer to this question is the framework for semantics-supported design of workflows that has been developed and challenged in the work underlying this this book with a particular focus on the bioinformatics application domain. Its major achievements and open challenges are described in the following.

9.1.1 Achievements

Figure 9.1 revisits the tradeoff between simplicity and generality of software frameworks of Figure 1.2, but additionally illustrates the role of workflow management systems in the software environments landscape. It distinguishes further between two kinds of workflow systems:

- *Domain-independent workflow systems* simplify software development by abstracting from programming details and by providing additional functionality for supporting, e.g., agile development, execution and deployment. These frameworks do usually not, however, address the application-specific customization of the workflow environment, so that the workflow

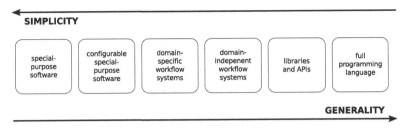

Fig. 9.1 Trade-off between simplicity and generality in software environments

design takes place on the level of abstraction defined by the framework, rather than on the user level.

- *Domain-specific workflow systems* focus on user-level service composition making use of a specific (semantic) domain model, which allows the user to work with the services and workflows using the technical language of his domain. These frameworks often also apply semantics-based methods for (semi-) automatic workflow composition. They are, however, usually restricted to the services covered by the domain model.

According to this classification scheme, Bio-jETI is both a domain-independent and a domain-specific workflow management system: The jABC framework with its broad range of established plugins provides a mature (domain-independent) workflow management system, and the loose programming approach implemented by the PROPHETS plugin allows for the incorporation of (domain-specific) semantic support on demand. In fact, the workflow development styles complement each other rather than having to be applied mutually exclusive. Thus, Bio-jETI provides a maximum balance of simplicity and generality. This acknowledges the observation that convenience for the user can not be achieved without specializing the software, but at the same time shows that domain-specialization does not have to impact the flexibility that is supported by a workflow framework.

In greater detail, Bio-jETI is a framework for service-oriented, model-driven design, execution, verification and deployment of bioinformatics workflows, extended by functionality for semantics-based, (semi-) automatic workflow composition. It is unique in its application of different flavors of formal methods, which facilitate constraint-guarded and constraint-driven workflow design:

- *Constraint-guarded workflow design* means that workflow design is accompanied by the continuous evaluation of workflow-level constraints, alerting the workflow designer when constraints are violated. In Bio-jETI, the GEAR model checking plugin is used to monitor workflow design in terms of constraints that are expressed in a temporal logic. Thus, in addition to the speedup of in silico experimentation and the abstraction from programming details that are also achieved by other commonly known systems for bioinformatics workflow design, constraint-guarded workflow design as in Bio-jETI provides means for reasoning about workflow properties at the user level.

- *Constraint-driven workflow design* extends the application of constraints further, systematically using them for high-level descriptions of individual components and entire workflows, which can then be translated automatically into concrete workflows that conform to the constraints by design. The PROPHETS plugin supports the synthesis of workflows based on specifications in terms of temporal-logic constraints according to the loose programming paradigm. In particular, it allows users to describe the intended workflow at an abstract, less technical level by providing simple means for sketching workflow "skeletons" and for expressing workflow constraints in terms of domain-specific vocabularies. It is applied for (semi-) automatic workflow design within Bio-jETI.

Bio-jETI's workflow design approach has been applied to a number of academic and real-life bioinformatics workflow scenarios. The application examples presented in this book cover different thematic areas (phylogenetic analyses, PCR primer design, metabolic flux analysis, and microarray data analysis), different software components (commonly known tool collections as well as special-purpose services), different service technologies (standard Web Services as well as especially created jETI services), and also workflows of different complexity. Their different characteristics allowed for the assessment of the current implementation in particular and provided insights regarding the capabilities and limitations of the approach in general.

Based on the experiences gained by working on the application scenarios and the results of the evaluation, a set of *loose programming pragmatics* has been formulated that provide general guidelines for adequate domain modeling and synthesis application. With regard to domain modeling, they emphasize the importance of adequate granularity of service and data types, and of a precise domain-specific vocabulary and clear semantic service interface descriptions in the domain models. With regard to the actual workflow design, they advocate an incremental exploration of the solution space, until the intended or an adequate set of solutions is obtained.

9.1.2 Remaining Challenges

There is of course room for improvement of the current implementation of the framework, but for many issues it is already relatively clear how they can be addressed, so that solving them is more or less a matter of doing (cf. Section 9.2). However, for two open issues that are crucial for the future success of the method, namely the performance of the synthesis algorithm and the origination of "good" domain models, the answers are not as clear, so that two major challenges remain.

Synthesis Performance

As detailed in Section 7.1.2, the performance of the synthesis algorithm suffers from the state explosion problem [324]: the size of the synthesis universe, and thus the time required for the (breadth-first) search of solutions in the universe, grows exponentially with the number of services in the domain model. Section 7.1.2 also shows, however, that despite state explosion, there is potential for improving the performance of the synthesis' search, as, for instance, constraints can not only reduce the number of solutions effectively, but also size of the search space.

At the downside, applying lots of constraints impacts the performance of the construction of the formula automaton from the constraints, which is done prior to the actual search. Although this step is usually faster and hence not crucial for the overall execution time of the synthesis, there is clear potential for improvement that should be exploited.

Domain Modeling

One central issue of all automatic workflow composition approaches is that considerable effort is required to set up adequate domain models. The quality of the obtained solutions crucially depends on the quality of the available information about the services of the domain. Merely syntactic information (as provided by most programming language APIs or standard web service interfaces) is not sufficient. Proper semantic descriptions of services and data types are required to obtain meaningful results. For instance, assume a web service that returns a job ID, and another service that consumes a nucleotide sequence. From a programming language point of view, both would be classified as character sequences (strings) and would thus be assumed to match. While this is syntactically correct, submitting a job ID to a sequence analysis service would of course not lead to the desired results, so a more precise, semantic interface description is required. Furthermore, adequate domain modeling requires to incorporate as much domain knowledge as possible, far beyond the mere technical aspects of the different types and services in order to enable efficient working with large, heterogeneous and hence manually intractable service collections.

While it is easy to see that the domain modeling is the key of making information technology available to users, as it can enable them to express their problems in their own language terms, finding or defining semantically appropriate service and type descriptions is indeed a difficult task [190], and it is not clear where the manifold ingredients for a good domain model should ideally come from. Basic domain-specific vocabulary can be provided by ontologies like EDAM (cf. Section 3.3.1), but refinements or additions may be necessary to cover the terminology of specific applications. Service annotations should ideally come directly from their providers (cf. [127]), but as the services themselves can be automatically imported or retrieved from component repositories as well as specifically implemented by a programmer, broadly accepted conventions for service annotations would be advantageous. Additional domain knowledge, such as constraints that describe frequently occurring workflow patterns or other pieces of specific expert knowledge, could as well be shared via public repositories or defined by the individual user according to the requirements of his particular application.

Summarizing, the future success of semantics-based automatic workflow composition largely depends on the availability of ready-to-use, adequate domain models. Although it is relatively clear how they should be built from a technical point of view, several practical questions regarding how to achieve this at a large, community-attracting scale are unanswered.

Note at this point that the services in the application domains considered in the scope of this work are characterized by being typically *stateless* (cf. [342]), that is, they exhibit the same behavior every time they are called. For such services and their compositions, it is comparatively easy to properly define the semantics along the lines discussed in Section 7.2.2, and hence to apply constraint-driven workflow synthesis method as described in this book. In contrast, in application domains with *stateful* services (i.e., services whose behavior can depend on their own execution history or other dynamically changing conditions), definition of semantics is inherently more difficult, and accordingly also other methods are needed for automatic service composition. This is, for instance, specifically addressed in the scope of the EU project CONNECT [139, 3].

9.2 Outlook

The major contributions of this work lie in the application of constraint-based methods to a comprehensive real-world application domain in order to achieve user-level workflow development, and the discussion of the corresponding achievements, open issues and experiences. This line of research can be continued in three major directions, which are described in greater detail in the following Sections 9.2.1, 9.2.2 and 9.2.3, respectively: the further development of the workflow framework, the continuation and enhancement of the bioinformatics applications, and the transfer of the methodology to other application domains.

9.2.1   Further Development of the Workflow Framework

Several results and experiences from the work described in this book concern the jABC framework and associated technologies. They have already influenced the development of the currently available framework, but moreover provide valuable input for the design and implementation of its currently emerging successor. Especially the ideas for further development of

the PROPHETS synthesis framework and the introduction of explicit variability management, which are described in greater detail in the following,
can be regarded as direct outcomes of this work.

Synthesis Framework

This book has demonstrated that the PROPHETS plugin, which is the current reference implementation of the loose programming paradigm, facilitates
semantics-based, (semi-) automatic workflow design within the jABC framework for service-oriented, model-driven process/workflow development. Open
issues involve in particular the performance of the current implementation of
the synthesis algorithm, the flexibility of the specifications, and an instance-
based synthesis procedure, but also different usability aspects ought to be
addressed in the scope of future work. The major ideas in these regards are
sketched in the following.

Performance

Improving synthesis performance can essentially be done by improving the
performance of the automaton construction and of the search for solutions (cf.
Section 9.1.2). As for the construction of the automaton, existing ideas comprise an adaption of the fast LTL to Büchi Automata Translation described
in [101] for finite automata, or a direct compositional automaton construction strategy, for which ideas like Second-order Value Numbering [213] could
be applied.

Regarding the actual search for solutions, the main idea for improving
the performance of the synthesis algorithm is to make use of domain-specific
heuristics [255, Section 3.6] for the search for solutions. As discussed earlier,
constraints can reduce the size of the search space effectively, especially when
expressing (domain-) specific knowledge about the *intended* solutions. Thus,
domain-specific search heuristics might exploit domain-specific knowledge,
such as typical workflow patterns, to better recognize adequate solutions and
thus speed up the search.

The impact of state explosion, however, remains also with improvements
as sketched above. Hence, additional, pragmatic strategies are needed for
making synthesis feasible also for average desktop or laptop computers. The
simplest idea in this regard is to execute the synthesis remotely on a truly
performant server. More sophisticated would be to apply a kind of divide-
and-conquer strategy (cf. Section 7.2.3): Instead of synthesizing one long
sequence of services, several smaller sub-workflows are synthesized and then
strung together. Due to the exponential runtime behavior of the synthesis,
the decrease of the maximal search depth to be explored would then result in
a better overall execution time. The central challenge here is how to split the
initial synthesis into adequate, definite sub-problems. One possible starting
point could be to identify elements in the solution that are guaranteed by

the applied constraints, such as a service whose existence in the solution is enforced by constraints, so that one synthesis run can be performed to generate the workflow up to this service, and another for creating the rest of the workflow.

Specification flexibility

A central functionality improvement in order to increase the flexibility of the synthesis method would be to distinguish more clearly between domain- and problem-specific constraints. Currently, all constraints are treated equally and applied to the same scope (i.e., the currently considered loosely specified branch). The only difference between domain- and problem-specific constraints is that the former are defined along with the domain model, while the latter are dynamically added during the synthesis process. Actually, however, it is desirable for the domain constraints to be applied globally to the entire workflow models, while problem-specific constraints may only be associated to a particular loosely specified branch. This goes in line with the ultimate ambition of loose programming [178]: finally, the whole workflow model as defined by the user will be treated as one comprehensive loose specification. The synthesis framework detects the actually underspecified parts automatically, and automatically creates a fully specified, readily executable model corresponding to the abstract description.

In this regard also relevant is the appropriate treatment of second order effects, which may occur when the workflow model contains several loose branches: In general, the synthesis algorithm can not treat multiple loose branches independently, since the concretization of one may cause side effects that influence others (positively or negatively). An approach that resolves the (potentially) occurring side-effects following a strategy based on property-oriented expansion [296] is currently being developed.

Instance-based synthesis

The current implementation of the loose programming framework performs the synthesis based on data *types*, rather than on concrete *instances* in terms of the services' input/output parameters. Consequently, if several parameters of the same type are involved, the synthesis framework is not able to tell them apart. As detailed in Section 2.3.2, this can cause ambiguities, especially with respect to inserting and instantiating services that have more than one parameter of a particular type. The simple heuristics for the parameter instantiation that is currently applied by PROPHETS does indeed result in correctly executable workflows in most cases studied. However, in particular cases the parameters of the SIBs had to be (re-) configured manually in order to obtain fully executable workflows.

To properly overcome these issues is a central aspect of future work on the framework. One possibility would be to develop more sophisticated heuristics for the instantiation, for instance by taking into account more of the available domain information or by recording and later reproducing the decisions taken by the synthesis algorithm during the assembly of the solutions. Ideal and more substantial, however, would be to enable the framework to directly perform instance-based workflow synthesis and thus avoiding ambiguities from the beginning.

Usability

In order to improve plugin usability, especially more sophisticated solution choosing mechanisms could be applied. A very effective feature would be the prioritization of solutions, for instance along the lines of [289], in order to achieve a more meaningful sorting. Furthermore, it is envisaged to replace the current functionality for choosing a solution from a list of textual representations by a graphical, interactive solution chooser that visualizes the entire set of solutions in a more comprehensible automaton representation, and supports different mechanisms for choosing a particular solution. An additional helpful feature in this regard would be kind of "automatic explanation" mechanism of the synthesis procedure itself. This would make the individual steps that are taken by the algorithm comprehensible through visualization of the current state of the search, possibly also allowing the user to influence the search procedure manually.

Minor usability improvements concern, for example, the further simplification of domain modeling, for instance by providing specific wizards for service descriptions and thus freeing the user completely from dealing with configuration files.

Explicit Variability Management

Model-based workflow development with the jABC framework and related technologies has always focused on supporting the agile handling of variant-rich software applications in different ways [208, 210]. Already the basic jABC framework provides a workflow definition layer at which variants of workflows can easily be built, which has been demonstrated for bioinformatics workflows in [172]. The abstract workflow descriptions that are applied by PROPHETS are furthermore declarative (but nevertheless intuitive) workflow specifications that represent all possible variations.

Unless explicitly configured by the user in a different way, the PROPHETS plugin makes the variant-richness of the workflows visible to the user: he can select a solution from a set of candidates, or he can decide to edit the applied constraints and run the synthesis again in order to refine the set of candidate solutions before selecting one manually. In fact, this is a striking difference to the other approaches that have been considered in the scope of this book (cf. Chapter 8): the selection of a particular solution is typically left to the composition algorithms, which simply take the first solution that is found, or apply more or less sophisticated heuristics for this task. Somehow, this is like oversimplifying the workflow composition problem by regarding it as a jigsaw puzzle, where the goal is to find the one possible combination of pieces. In realistic application domains for scientific workflow composition, however, there are different pieces (services) that fit next to each other, and consequently there are several possible combinations that are worth being considered – also at the user level.

However, currently neither the basic jABC nor the PROPHETS manage workflow variability explicitly, that is, workflow variants are more or less created on the fly, rather than being systematically handled and described. Consequently, it is desirable to incorporate explicit means for *variability management* [330] in order to support the development of variant-rich workflows better.

As variant-rich workflows can be regarded as a particular form of software product lines (SPLs) [253, 329], it suggests itself to pick up on the available work on variability management of SPLs, such as [227, 99, 77, 65], here. According to, e.g., [319, 169, 272, 143], two principal kinds of variability modeling can be distinguished:

1. *Structure-oriented* approaches apply the concept of *variation points* [253, Chapter 4], that is, explicitly defined points in the workflow model where services or parameters have several alternatives. The different variations that can realize the variation point are attached to it, and concrete workflows are built by selection of variants at the defined variation points.
2. *Behavior-oriented* approaches apply notions of labeled transition systems [149] for representing behavioral variability (cf., e.g., [33, 71]). Concrete workflow variants are derived by identifying the sub-systems of the model that conform to the associated constraints or conditions.

Conveniently, both kinds of variability management are already supported by the framework:

1. As detailed in [142, Section 4.1.4] and [143], the jABC's SIBs for building hierarchical models can easily be interpreted as variation points in the sense of structural variability modeling. The recently introduced *Second-Order Servification* mechanism [239] goes further and makes it possible to define variants of jABC SLGs via second-order parameterization, and to exchange services and even subprocesses dynamically at runtime.
2. The constraint-driven workflow composition approach of PROPHETS provides the means for a very liberal, declarative, and indirect way of handling behavioral workflow variability [169, 271, 143]: Constraints in terms of a domain model specify a set of similar workflows, that is, variants of an implicitly described, abstract workflow. The actual product variants are automatically generated from these specifications and are thus correct by construction [170].

Thus, adding explicit variability management support does in fact not require comprehensive new developments, as several notions are already inherent in the present framework. It will rather be a question of consolidation and consistent naming to make the next generation of the framework directly recognizable as a mature variability management framework.

9.2.2   Continuation of the Bioinformatics Applications

The bioinformatics workflow scenarios discussed in this book are only four examples from the plethora of possible applications. In fact, there are already several further Bio-jETI workflow applications originating from different contexts available. For instance, LC/MS analysis [158] and orthologuous ID retrieval [200, 128] workflows were realized in the scope of the first collaborative projects, and functional annotation or other sequence analysis workflows were implemented by various student projects. Furthermore, some (still experimental) projects explore the workflow integration of the Cytoscape network analysis and visualization platform [280] and the metabolic flux analysis software OpenFlux [259], or address the realization of the DDBJ workflows [4]. So far, however, these applications have not made use of the developed methods for constraint-driven workflow design. Accordingly, future work will certainly comprise to challenge and evaluate the constraint-driven workflow design approach on these (and further) bioinformatics application scenarios.

In general, two major challenges can be identified with regard to the application of the constraint-based workflow design methods in the bioinformatics domain: capturing domain knowledge and promoting the methodology in the target user community. They are described in greater detail in the following.

Systematic Capturing of Domain Knowledge

As discussed above, a good domain model is necessary for the successful application of (semi-) automatic workflow composition methods, but the crucial aspect is how an adequate domain model should be developed. Although this aspect is considered from the Bio-jETI perspective here, it is in the same way relevant for other semantic workflow management systems that follow similar approaches.

As for the domain-specific vocabulary, EDAM has proved itself being an adequate basis, and service annotations in terms of the controlled vocabulary it defines facilitate finding possible workflows [180]. In some cases, however, additional terminology may be needed to describe highly specialized services or data. What is more, the workflows that are principally *possible* according to the mere interface annotations are not necessarily also the actually *intended* solutions. Accordingly, additional knowledge may have to be included in the domain model or in a concrete workflow specification to constrain the search further to the actually intended solutions. This additional knowledge can be provided by two kinds of constraints:

- Constraints that express relationships between individual services or data types, such as that service A must always be followed by service B, or that a service C must not be called before service D.

- Constraints that describe the overall structure of a workflow, like requiring that the services X, Y and Z have to be included in exactly this order.

The first kind of constraints can on the one hand already be provided by the domain modeler or later by the workflow designer, capturing the specific user's expert knowledge about the types and services in the domain. On the other hand, such information could systematically be included in the service documentations of the providers, which would prevent the redundant identification and formalization of the respective information by different users.

The second kind of constraints can of course be defined directly by the workflow designer, who uses constraints to describe the overall structure of the intended workflow. However, it would be even more interesting to provide ready-to-use (constraint) libraries of domain-specific workflow patterns, that is, templates of frequently occurring bioinformatics service compositions. This goes beyond the general, domain-independent workflow patterns identified and described in [326]. Possible sources for the systematic, large-scale identification of such domain-specific patterns are, for instance:

- *Event logs* of major service providers (such as the European Bioinformatics Institute, EBI) or service portals (such as BioMoby [341]), in which process mining frameworks (like, e.g., ProM [337, 328, 51]) could find several patterns,
- *Workflow provenance data*, that is, metadata that helps managing the large amounts of data in the course of scientific experiments [267], which in contrast to the mere event logs allows for a better discrimination of the data dependencies [354], and
- The *has_input* and *has_output relations* that the EDAM ontology (cf. Section 3.3.1) defines for the *Operation* terms in the ontology. Simply regarding the *Operation* terms as services, the synthesis methods of Bio-jETI, for instance, might easily find service sequences in the abstract terms of the ontology.

In addition to being used as workflow constraints by domain modelers and workflow designers, domain-specific workflow patterns could also be useful for improving the performance of the search for adequate workflows, for instance by their application for domain-specific search heuristics (see above).

In the same way as EDAM is a publicly available constituent of bioinformatics domain models, additional domain-specific knowledge as described above should ideally also be made available via a common public repository. Such a library of constraints that can be added and removed dynamically during workflow development would enable users to work and experiment with the domain in a very flexible manner. The details of such a platform would have to be negotiated within the community, but existing Semantic Web technology could certainly be employed for this purpose.

Promotion

A central question with regard to making third-party users use the framework is where and how in silico researchers would use semantics-based, automatic workflow composition functionality. Larger service collections like EMBOSS (cf. Section 3.3) or the BioCatalogue [41], which provide hundreds or even thousands of services, are in fact not manageable without systematic discovery or service composition techniques. When dealing with small domains, human experts may be unbeatable in composing tailored workflows, but not every human is an expert in bioinformatics services and data types, and also experts cannot always keep track of all changes in large domain libraries. Thus both experts and average users profit from tools that systematically collect and exploit semantic service annotations and appropriately formalized domain knowledge.

However, although there are many promising approaches to semantics-based, (semi-) automatic service composition, they have apparently not become widely established so far. One reason for this might be that biologists are reluctant to experiment with the new technologies, while bioinformaticians have accustomed themselves to scripting and conventional programming in a multitude of languages. Or, as stated by Ian Holmes in the BioWiki, "Bioinformatics workflows can be approached in several ways. Unix hackers often resort to GNU make, while computer scientists dream of more elegant approaches." [131]. Thus, computer science will have to explore further on how to make their more elegant and more powerful approaches to (semi-) automatic workflow design accessible to the broad biological user community.

The usability of concrete implementations of (semi-) automatic workflow composition in Bio-jETI (and other systems) could, for instance, be assessed by an evaluation framework as applied in [313, 244]. This would mean to carry out the usability evaluation based on analysis of actually created workflows, user questionnaires and semi-structured interviews. However, the mere technical aspects of usability are not the only aspects that are relevant for the success of a particular software system or methodology. In fact, it also largely depends on "the realms of sociology and psychology" [128], which are sometimes difficult to understand. Hence, it would be interesting to explore potential users' obstacles for using semantic systems more generally. It is likely that people simply mistrust the formal and/or semantic methods that are still very unfamiliar to them. Here, specific promotion and training (educational material, comprehensive tutorials, etc.) is required to get people accustomed to a new way of thinking about workflows.

Furthermore, experience shows that an entire software design *method* is difficult to explain and promote, since application experts tend to focus on the aspects related to their specific domain, expecting a concrete software *tool* for a very particular purpose. Thus, it seems to be a promising strategy to carefully prepare domain models for specific bioinformatics applications, and then release Bio-jETI together with a particular domain model as a custom-tailored, special-purpose workflow tool.

9.2.3   Transfer to other Application Domains

From an abstract informatics point of view, this work can be regarded as a case-oriented analysis and subsequent customization of a family of technologies, focusing on how the jABC modeling framework and associated technologies can be brought into application in a particular application domain. Other examples of comprehensive jABC applications are, for instance, the Genesys framework for the construction of code generators developed by Sven Jörges (and described in detail in his dissertation [142]), the LearnLib Studio [260, 220, 219] for the design and management of automata learning processes, the OCS conference system [238], the applications for business process management by Markus Doedt [87, 199], and the project management applications developed by Steve Boßelmann [283].

The strong focus on semantics and constraints that is central for the bioinformatics workflow management methodology described in this book, however, is so far unique among the larger jABC applications. As they have proven themselves successful, it will be interesting to challenge the developed methods for constraint-driven workflow management also in the scope of other existing jABC applications and in further, especially scientific, application domains. As the methods presented in this book have been developed with a focus on the bioinformatics application domain, it is in fact likely that they will especially be adequate for use in other scientific disciplines, which often have a similar technical infrastructure and technically similar data analysis processes. Tuning our constraint-based methods for use in a particular application domains inevitably leads to a better understanding of the domain characteristics: which components might be used for a certain task, which causalities need to be considered, and which incompatibilities constrain the variation potential are some examples of relevant knowledge. All this can be profitably used for the identification and adequate formalization of the domain knowledge that is actually critical for any concrete application.

Concretely, there have already been first considerations about applying the constraint-driven workflow design approach in the fields of computer linguistics and geoinformatics (particularly geographic information systems, GIS) data analysis processes:

- The computer linguistics community only begins to adopt service-oriented thinking and to provide software components for flexible re-use. First example applications (such as, e.g., the pipeline for computational historical linguistics described in [308]) show that workflows can contribute to an adequate management of variant-rich, complex analysis processes also in this field.
- For GIS applications, there is a plethora of tools, data and services available (cf., e.g., [316, 222]), constituting a service infrastructure similar to that of the bioinformatics domain. Likewise, the concept of using workflow methods for realizing GIS processes grows in popularity (cf., e.g., [191, 352]), and being aware of the importance of user-level access to the

involved entities (cf., e.g.,[68]) also the development of specific domain ontologies has already started (cf., e.g., [333, 194]).

The application scenarios described in this book are local in the sense the they deal with particular data analysis workflows, such as the design of PCR primers (cf. Chapter 5) or the analysis of data from metabolic flux experiments (cf. Chapter 4). They are, however, usually parts of larger research processes: Complex genomics and expression profiling experiments, for instance, require the amplification of DNA fragments and thus suitable primers for the PCR at different points. Or, as another example, the development of diabetes treatments needs detailed knowledge about cell metabolism (which can be provided by the flux experiments), but also involves different other fields of work, such as the actual design of pharmaceuticals, the clinical trials and the drug licensing procedures.

This suggests to extend the application scope of the developed methodologies from the currently addressed local data analysis workflows to global research management processes. These larger, typically hierarchical processes can as well be modeled with the framework in order to optimize the research work from a more global perspective. This may e.g. comprise the automatic comparison with previous analysis results, the generation of reports, as well as an adequate archiving of the results for later reuse.

On the one hand, this is facilitated by the flexibility of the framework with regard to the concrete applications. In fact, the characteristics of the individual applications are essentially captured by the respective domain models. Hence, the methods can be expected to apply also to the management of the higher-level research management processes, which do in fact have a business-process flavor rather than the typical characteristics of scientific application domains. On the other hand, the hierarchical control-flow structure of the workflow and process models is central with regard to achieving this goal. Hierarchical models are required to structure complex processes adequately and represent them at different levels of detail/granularity, according to the focus of the respective user. The control-flow structure of the models allows for a more decisive representation of process flows than data-flow modeling could achieve, and by being less focused on the data, it is also more suitable for modeling the higher-level processes of complex research projects.

Finally, it is especially the application of a coherent modeling formalism throughout the framework which will facilitate the adoption of the methodology by the different users. In combination with the specific domain models that allow the users to use the terms of their respective technical languages, they will enable the users to effectively work at the various areas of expertise and levels of granularity. Thus, true user-level handling of all local workflows and global processes involved in complex research projects will become possible.

References

1. Bioconductor - open source software for bioinformatics,
   `http://www.bioconductor.org/` (last accessed May 5, 2013)
2. BioMake and Skam project homepage, `http://skam.sourceforge.net/` (last accessed May 5, 2013)
3. CONNECT - Emergent Connectors for Eternal Software Intensive Networked Systems, `https://www.connect-forever.eu/index.html` (last accessed May 5, 2013)
4. DDBJ Web API for Biology, `http://xml.nig.ac.jp/workflow/` (temporarily suspended since February 15, 2012)
5. EMBRACE (European Model for Bioinformatics Research and Community Education) Network of Excellence, `http://www.embracegrid.info` (last accessed May 5, 2013)
6. EMBRACE Ontology for Data and Methods (EDAM),
   `http://edamontology.sourceforge.net/` (last accessed May 5, 2013)
7. ETH - JOpera for Eclipse - Welcome to the JOpera Project,
   `http://www.jopera.ethz.ch` (last accessed May 5, 2013)
8. GNU make - GNU project - Free Software Foundation,
   `http://www.gnu.org/software/make/` (last accessed May 5, 2013)
9. Google Search, `http://www.google.com/` (last accessed May 5, 2013)
10. ImageMagick: Convert, Edit, Or Compose Bitmap Images,
    `http://www.imagemagick.org/` (last accessed May 5, 2013)
11. Kepler Analytical Component Repository,
    `http://library.kepler-project.org/kepler/style/skins/kepler/`
    (last accessed May 5, 2013)
12. Pipes: Rewire the web, `http://pipes.yahoo.com/pipes/` (last accessed May 5, 2013)
13. Ruby Programming Language, `http://www.ruby-lang.org/` (last accessed May 5, 2013)
14. Soaplab, `http://soaplab.sourceforge.net/soaplab1/` (last accessed May 5, 2013)
15. Soaplab2, `http://soaplab.sourceforge.net/soaplab2/` (last accessed May 5, 2013)
16. The Comprehensive R Archive Network, `http://cran.r-project.org/` (last accessed May 5, 2013)

17. The Perl Programming Language, http://www.perl.org/ (last accessed May 5, 2013)

18. The R Project for Statistical Computing, http://www.r-project.org/ (last accessed May 5, 2013)

19. Unreal Engine 4 - Feature-Präsentation zu Epics neuer Grafik-Engine - Video bei Gamestar.de, http://www.gamestar.de/index.cfm?pid=1589&pk=66370 (last accessed May 5, 2013)

20. Introduction to UDDI: Important features and functional concepts (October 2004), http://www.uddi.org/pubs/uddi-tech-wp.pdf (last accessed May 5, 2013)

21. E. coli outbreak: New Comprehensive Comparisons, Pathosystem Resource Integration Center (PATRIC), eNews (June 2011), http://enews. patricbrc.org/1172/e-coli-outbreak-new-comprehensive-comparisons/ (last accessed May 5, 2013)

22. Affymetrix, Inc.: Affymetrix CEL Data File Format (2009), http://www.affymetrix.com/support/developer/powertools/changelog/ gcos-agcc/cel.html (last accessed May 5, 2013)

23. Aho, A.V., Lam, M.S., Sethi, R., Ullman, J.D.: Compilers: Principles, Techniques, and Tools, 2nd edn. Addison Wesley (2007)

24. Allen, P.: Service Orientation: Winning Strategies and Best Practices. Cambridge University Press, New York (2006)

25. Alonso, G.: BioOpera: Grid Computing in Virtual Laboratories. ERCIM News (45) (April 2001)

26. Alonso, G., Pautasso, C.: JOpera: a Toolkit for Efficient Visual Composition of Web Services. International Journal of Electronic Commerce (IJEC) 9(2), 107–141 (2004)

27. Altintas, I., Barney, O., Cheng, Z., et al.: Accelerating the scientific exploration process with scientific workflows. Journal of Physics: Conference Series 46(1), 468 (2006)

28. Altintas, I., Berkley, C., Jaeger, E., et al.: Kepler: An Extensible System for Design and Execution of Scientific Workflows. In: Proceedings of the 16th International Conference on Scientific and Statistical Database Management (SSDBM 2004), pp. 21–23. IEEE Computer Society (June 2004)

29. Altschul, S.F., Gish, W., Miller, W., Myers, E.W., Lipman, D.J.: Basic local alignment search tool. Journal of Molecular Biology 215(3), 403–410 (1990)

30. Andrews, T., Curbera, F., Dholakia, H., et al.: Business Process Execution Language for Web Services Version 1.1 (May 2003), http://msdn.microsoft.com/en-US/library/ee251594(v=bts.10).aspx (last accessed May 5, 2013)

31. Arsanjani, A., Booch, G., Boubez, T., et al.: SOA Manifesto (October 2009), http://www.soa-manifesto.org/default.html (last accessed May 5, 2013)

32. Ashburner, M., Ball, C.A., Blake, J.A., et al.: Gene ontology: tool for the unification of biology. The Gene Ontology Consortium. Nature Genetics 25(1), 25–29 (2000)

33. Asirelli, P., ter Beek, M.H., Fantechi, A., Gnesi, S.: A logical framework to deal with variability. In: Méry, D., Merz, S. (eds.) IFM 2010. LNCS, vol. 6396, pp. 43–58. Springer, Heidelberg (2010)

34. Bakera, M., Jörges, S., Margaria, T.: Test your Strategy: Graphical Construction of Strategies for Connect-Four. In: Proceedings of the 2009 14th IEEE International Conference on Engineering of Complex Computer Systems, ICECCS 2009, pp. 172–181. IEEE Computer Society, Washington, DC (2009)

35. Bakera, M., Margaria, T., Renner, C., Steffen, B.: Verification, Diagnosis and Adaptation: Tool-supported enhancement of the model-driven verification process. In: Revue des Nouvelles Technologies de l'Information (RNTI-SM-1), pp. 85–98 (December 2007)

36. Bakera, M., Margaria, T., Renner, C., Steffen, B.: Tool-supported enhancement of diagnosis in model-driven verification. Innovations in Systems and Software Engineering 5, 211–228 (2009)

37. Barker, A., van Hemert, J.: Scientific Workflow: A Survey and Research Directions. In: Wyrzykowski, R., Dongarra, J., Karczewski, K., Wasniewski, J. (eds.) PPAM 2007. LNCS, vol. 4967, pp. 746–753. Springer, Heidelberg (2008)

38. Bausch, W., Pautasso, C., Schaeppi, R., Alonso, G.: BioOpera: Cluster-aware Computing. In: Proceedings of the 4th IEEE International Conference on Cluster Computing, pp. 99–106 (2002)

39. Beck, K., Andres, C.: Extreme programming explained: embrace change. Addison-Wesley Professional (2004)

40. Berners-Lee, T., Hendler, J., Lassila, O.: The Semantic Web - A new form of Web content that is meaningful to computers will unleash a revolution of new possibilities. Scientific American 284(5), 34–43 (2001)

41. Bhagat, J., Tanoh, F., Nzuobontane, E., et al.: BioCatalogue: a universal catalogue of web services for the life sciences. Nucleic Acids Research 38(suppl. 2), W689–W694 (2010)

42. Bille, S.: Vergleich und Bewertung von Bioinformatik-Workflow-Systemen auf Basis der DDBJ-Workflows. Diploma thesis, Technische Universität Dortmund, Fakultät für Informatik, Lehrstuhl für Programmiersysteme (2010)

43. Blank, L.M., Sauer, U.: TCA cycle activity in Saccharomyces cerevisiae is a function of the environmentally determined specific growth and glucose uptake rates. Microbiology 150(pt. 4), 1085–1093 (2004)

44. Blankenberg, D., Gordon, A., Von Kuster, G., et al.: Manipulation of FASTQ data with Galaxy. Bioinformatics 26(14), 1783–1785 (2010)

45. Blankenberg, D., Von Kuster, G., Coraor, N., et al.: Galaxy: a web-based genome analysis tool for experimentalists.. In: Ausubel, F.M., et al. (eds.) Current Protocols in Molecular Biology, ch. 19. John Wiley & Sons, Inc. (January 2010)

46. Bock, C., Kuster, G., Halachev, K., et al.: Web-Based Analysis of (Epi-) Genome Data Using EpiGRAPH and Galaxy. In: Barnes, M.R., Breen, G. (eds.) Genetic Variation. Methods In Molecular Biology^{TM}, vol. 628, pp. 275–296. Humana Press (2010)

47. Bolstad, B.M.: Low Level Analysis of High-density Oligonucleotide Array Data: Background, Normalization and Summarization. Ph.D. thesis, University of California, Berkeley (2004)

48. Bolstad, B.M., Brettschneider, J., Buhlmann, P., et al.: Bioinformatics and Computational Biology Solutions Using R and Bioconductor (Statistics for Biology and Health). Springer-Verlag New York, Inc., Secaucus (2005)

49. Booth, D., Haas, H., McCabe, F., et al.: Web Services Architecture. W3C Working Group Note (February 2004), http://www.w3.org/TR/ws-arch/ (last accessed May 5, 2013)

50. Borner, J.: A molecular approach to chelicerate phylogeny. Diploma thesis, Universität Hamburg (2010)

51. Bose, R.P.J.C., van der Aalst, W.M.P.: Abstractions in Process Mining: A Taxonomy of Patterns. In: Dayal, U., Eder, J., Koehler, J., Reijers, H.A. (eds.) BPM 2009. LNCS, vol. 5701, pp. 159–175. Springer, Heidelberg (2009)

52. Bowers, S., Ludaescher, B., Ngu, A., Critchlow, T.: Structured Composition of Dataflow and Control-Flow for Reusable and Robust Scientific Workflows. In: Symposium on Applied Computing (2006)

53. Bowers, S., McPhillips, T., Riddle, S., Anand, M.K., Ludäscher, B.: Kepler/p-POD: Scientific Workflow and Provenance Support for Assembling the Tree of Life. In: Freire, J., Koop, D., Moreau, L. (eds.) IPAW 2008. LNCS, vol. 5272, pp. 70–77. Springer, Heidelberg (2008)

54. Brin, S., Page, L.: The anatomy of a large-scale hypertextual Web search engine. Computer Networks and ISDN Systems 30(1-7), 107–117 (1998)

55. Bryson, B.: A Short History of Nearly Everything. Black Swan (2004)

56. Buetow, K., Klausner, R., Fine, H., et al.: Cancer Molecular Analysis Project: weaving a rich cancer research tapestry. Cancer Cell 1(4), 315–318 (2002)

57. Burger, A., Paschke, A., Romano, P., Splendiani, A.: Semantic Web Applications and Tools for Life Sciences 2008, Proc. of 1st Workshop SWAT4LS 2008, Edinburgh, United Kingdom. CEUR Workshop Proceedings (November 2008)

58. Burkart, O., Steffen, B.: Model checking for context-free processes. In: Cleaveland, W.R. (ed.) CONCUR 1992. LNCS, vol. 630, pp. 123–137. Springer, Heidelberg (1992)

59. Burkart, O., Steffen, B.: Composition, decomposition and model checking of pushdown processes. Nordic J. of Computing 2(2), 89–125 (1995)

60. Burkart, O., Steffen, B.: Model Checking the Full Modal Mu-Calculus for Infinite Sequential Processes. In: Degano, P., Gorrieri, R., Marchetti-Spaccamela, A. (eds.) ICALP 1997. LNCS, vol. 1256, pp. 419–429. Springer, Heidelberg (1997)

61. Burnett, M.M., Scaffidi, C.: End-User Development. In: Soegaard, M., Dam, R.F. (eds.) Encyclopedia of Human-Computer Interaction, ch. 10. Interaction-Design.org (2012), http://www.interaction-design.org/encyclopedia/end-user_development.html

62. Cardoso, J., van der Aalst, W. (eds.): Handbook of Research on Business Process Modeling. Information Science Reference - Imprint of: IGI Publishing, Hershey (2009)

63. Carlson, M., Falcon, S., Pages, H., Li, N.: hgu95av2.db: Affymetrix Human Genome U95 Set annotation data, chip hgu95av2 (2011), http://bioc.ism.ac.jp/2.6/data/annotation/manuals/hgu95av2.db/man/hgu95av2.db.pdf (last accessed June 25, 2012)

64. Carrere, S., Gouzy, J.: REMORA: a pilot in the ocean of BioMoby web-services. Bioinformatics 22(7), 900–901 (2006)

65. Chen, L., Babar, M.A.: A systematic review of evaluation of variability management approaches in software product lines. Information and Software Technology 53(4), 344–362 (2011)

66. Chen, L., Shadbolt, N.R., Goble, C., Tao, F., Cox, S.J., Puleston, C., Smart, P.R.: Towards a Knowledge-Based Approach to Semantic Service Composition.

In: Fensel, D., Sycara, K., Mylopoulos, J. (eds.) ISWC 2003. LNCS, vol. 2870, pp. 319–334. Springer, Heidelberg (2003)

67. Cheyer, A., Martin, D.: The Open Agent Architecture. Journal of Autonomous Agents and Multi-Agent Systems 4(1), 143–148 (2001)

68. Chin, G., Sivaramakrishnan, C., Critchlow, T., Schuchardt, K., Ngu, A.H.: Scientist-Centered Workflow Abstractions via Generic Actors, Workflow Templates, and Context-Awareness for Groundwater Modeling and Analysis. In: 7th IEEE World Congress on Services (SERVICES 2011), pp. 176–183. IEEE Computer Society, Los Alamitos (2011)

69. Christensen, E., Curbera, F., Meredith, G., Weerawarana, S.: Web Services Description Language (WSDL) 1.1. W3C Note (March 2001), http://www.w3.org/TR/wsdl.html (last accessed May 5, 2013)

70. Clarke, E.M., Grumberg, O., Peled, D.A.: Model Checking. The MIT Press (1999)

71. Classen, A., Heymans, P., Schobbens, P.-Y., Legay, A., Raskin, J.-F.: Model checking lots of systems: efficient verification of temporal properties in software product lines. In: Proceedings of the 32nd ACM/IEEE International Conference on Software Engineering, ICSE 2010, vol. 1, pp. 335–344. ACM, New York (2010)

72. Cochrane, G., Karsch-Mizrachi, I., Nakamura, Y.: On behalf of the International Nucleotide Sequence Database Collaboration: The International Nucleotide Sequence Database Collaboration. Nucleic Acids Research 39(Database), D15–D18 (2010)

73. Cope, L.M., Irizarry, R.A., Jaffee, H.A., Wu, Z., Speed, T.P.: A benchmark for Affymetrix GeneChip expression measures. Bioinformatics 20(3), 323–331 (2004)

74. Costabile, M.F., Dittrich, Y., Fischer, G., Piccinno, A.: IS-EUD 2011. LNCS, vol. 6654. Springer, Heidelberg (2011)

75. Curcin, V., Ghanem, M., Guo, Y.: Analysing scientific workflows with Computational Tree Logic. Cluster Computing 12, 399–419 (2009)

76. Curcin, V., Ghanem, M., Guo, Y., et al.: IT Service Infrastructure for Integrative Systems Biology. In: Proceedings of the 2004 IEEE International Conference on Services Computing, pp. 123–131. IEEE Computer Society, Washington, DC (2004)

77. Czarnecki, K.: Variability Modeling: State of the Art and Future Directions. In: VaMoS, p. 11. ICB-Research Report No. 37, University of Duisburg Essen (2010)

78. Dall'Olio, G.M.: GNU/make and bioinformatics. Presentation at BioEvo technical seminars (February 2009), http://www.slideshare.net/giovanni/makefiles-bioinfo (last accessed May 5, 2013)

79. Dalman, T., Droste, P., Weitzel, M., Wiechert, W., Nöh, K.: Workflows for Metabolic Flux Analysis: Data Integration and Human Interaction. In: Margaria, T., Steffen, B. (eds.) ISoLA 2010, Part I. LNCS, vol. 6415, pp. 261–275. Springer, Heidelberg (2010)

80. Darwin, C.: On the Origin of Species by Means of Natural Selection, or the Preservation of Favoured Races in the Struggle for Life. John Murray, London (1859)

81. Dayhoff, M., Schwartz, R., Orcutt, B.: A model of evolutionary change in proteins. Atlas of Protein Sequence and Structure 5(suppl. 3), 345–352 (1978)

82. DeBellis, M., Haapala, C.: User-centric Software Engineering. IEEE Expert 10(1), 34–41 (1995)
83. Deelman, E., Singh, G., Hui Su, M., et al.: Pegasus: a framework for mapping complex scientific workflows onto distributed systems. Scientific Programming Journal 13, 219–237 (2005)
84. DiBernardo, M., Pottinger, R., Wilkinson, M.: Semi-automatic web service composition for the life sciences using the BioMoby semantic web framework. Journal of Biomedical Informatics 41(5), 837–847 (2008)
85. Dijkstra, E.W.: A note on two problems in connection with graphs. Numerische Mathematik 1, 269–271 (1959)
86. Doedt, M.: Erweiterung der jABC-Framework-Bibliothek um eine modular anpassbare Ausführungsschicht. Diploma thesis, Universität Dortmund (2006)
87. Doedt, M., Steffen, B.: Requirement-Driven Evaluation of Remote ERP-System Solutions: A Service-oriented Perspective. In: 34th IEEE Software Engineering Workshop (SEW 2011), pp. 57–66 (June 2011)
88. Droste, P., von Lieres, E., Wiechert, W., Nöh, K.: Customizable Visualization on Demand for Hierarchically Organized Information in Biochemical Networks. In: Barneva, R.P., Brimkov, V.E., Hauptman, H.A., Natal Jorge, R.M., Tavares, J.M.R.S. (eds.) CompIMAGE 2010. LNCS, vol. 6026, pp. 163–174. Springer, Heidelberg (2010)
89. Ebert, B.E.: A systems approach to understand and engineer whole-cell redox biocatalysts. Dissertation, Fakultät Bio- und Chemieingenieurwesen, Technische Universität Dortmund (2011)
90. Ebert, B.E., Lamprecht, A.-L., Steffen, B., Blank, L.M.: Flux-P: Automating Metabolic Flux Analysis. Metabolites 2(4), 872–890 (2012)
91. Eker, J., Janneck, J., Lee, E., et al.: Taming heterogeneity - the Ptolemy approach. Proceedings of the IEEE 91(1), 127–144 (2003)
92. Facchiano, A., Romano, P. (eds.): Proceedings of the Fifth International Workshop NETTAB 2005 on Workflows Management: New Abilities for the Biological Information Overflow (October 2005)
93. Falcon, S., Morgan, M., Gentleman, R.: An Introduction to Bioconductor's ExpressionSet Class (February 2007),
http://www.bioconductor.org/packages/2.9/bioc/vignettes/Biobase/inst/doc/ExpressionSetIntroduction.pdf (last accessed May 5, 2013)
94. Fischer, E., Sauer, U.: Metabolic flux profiling of Escherichia coli mutants in central carbon metabolism using GC-MS. European Journal of Biochemistry/FEBS 270(5), 880–891 (2003)
95. Fischer, E., Zamboni, N., Sauer, U.: High-throughput metabolic flux analysis based on gas chromatography-mass spectrometry derived 13C constraints. Analytical Biochemistry 325(2), 308–316 (2004)
96. Fowler, M., Parsons, R.: Domain-specific languages. Addison-Wesley/ACM Press (2011)
97. Freitag, B., Margaria, T., Steffen, B.: A Pragmatic Approach to Software Synthesis. In: Workshop on Interface Definition Languages, pp. 46–58 (1994)
98. Freitag, B., Steffen, B., Margaria, T., Zukowski, U.: An Approach to Intelligent Software Library Management. In: Proceedings of the 4th International Conference on Database Systems for Advanced Applications (DASFAA), pp. 71–78. World Scientific Press (1995)
99. Garces, K., Parra, C., Arboleda, H., Yie, A., Casallas, R.: Variability Management in a Model-Driven Software Product Line. Avances en Sistemas e Informática 4(2), 3–12 (2007)

100. Garvey, T.D., Lincoln, P., Pedersen, C.J., Martin, D., Johnson, M.: BioSPICE: access to the most current computational tools for biologists. Omics: A Journal of Integrative Biology 7(4), 411–420 (2003)

101. Gastin, P., Oddoux, D.: Fast LTL to Büchi Automata Translation. In: Berry, G., Comon, H., Finkel, A. (eds.) CAV 2001. LNCS, vol. 2102, pp. 53–65. Springer, Heidelberg (2001)

102. Gautier, L., Cope, L., Bolstad, B.M., Irizarry, R.A.: affy—analysis of Affymetrix GeneChip data at the probe level. Bioinformatics 20(3), 307–315 (2004)

103. Gentleman, R.: annotate: Annotation for microarrays (2010), http://www.bioconductor.org/packages/release/bioc/html/annotate.html (last accessed May 5, 2013)

104. Gentleman, R., Carey, V., Huber, W., Hahne, F.: genefilter: methods for filtering genes from microarray experiments (2010), http://bioc.ism.ac.jp/2.8/bioc/html/genefilter.html (last accessed May 5, 2013)

105. Gentleman, R.C., Carey, V.J., Bates, D.M., et al.: Bioconductor: open software development for computational biology and bioinformatics. Genome Biology 5(10), R80 (2004)

106. Ghanem, M., Curcin, V.: Scientific workflow systems - can one size fit all? In: Cairo International Biomedical Engineering Conference, CIBEC 2008, pp. 1–9 (2008)

107. Ghanem, M., Curcin, V., Wendel, P., Guo, Y.: Building and Using Analytical Workflows in Discovery Net, pp. 119–139. John Wiley & Sons, Ltd. (2009)

108. Ghanem, M., Guo, Y., Lodhi, H., Zhang, Y.: Automatic scientific text classification using local patterns: KDD CUP 2002 (task 1). ACM SIGKDD Explorations Newsletter 4, 95–96 (2002)

109. Giegerich, R., Meyer, F., Schleiermacher, C.: GeneFisher – software support for the detection of postulated genes. In: Proceedings of the International Conference on Intelligent Systems for Molecular Biology (ISMB), vol. 4, pp. 68–77 (1996)

110. Gil, Y.: From data to knowledge to discoveries: Artificial intelligence and scientific workflows. Science of Computer Programming 17(3), 231–246 (2009)

111. Gil, Y., Ratnakar, V., Deelman, E., Mehta, G., Kim, J.: Wings for Pegasus: creating large-scale scientific applications using semantic representations of computational workflows. In: Proceedings of the 19th National Conference on Innovative Applications of Artificial Intelligence, vol. 2, pp. 1767–1774. AAAI Press (2007)

112. Goble, C.A., Belhajjame, K., Tanoh, F., et al.: BioCatalogue: A Curated Web Service Registry for the Life Science Community. In: 3rd International Biocuration Conference, Nature Precedings, Berlin, April 16-18. Nature Publishing Group (April 2009)

113. Goble, C.A., Bhagat, J., Aleksejevs, S., et al.: myExperiment: a repository and social network for the sharing of bioinformatics workflows. Nucleic Acids Research 38(suppl. 2), W677–W682 (2010)

114. Goecks, J., Nekrutenko, A., Taylor, J., Team, T.G.: Galaxy: a comprehensive approach for supporting accessible, reproducible, and transparent computational research in the life sciences. Genome Biology 11(8), R86 (2010)

115. Goodman, N., Rozen, S., Stein, L.D.: Workflow Management Software for Genome-Laboratory Informatics, From a proposal to the National Institues of Health (1995)
116. Gordon, P., Sensen, C.: Seahawk: moving beyond HTML in Web-based bioinformatics analysis. BMC Bioinformatics 8(1), 208 (2007)
117. Görlach, K., Sonntag, M., Karastoyanova, D., Leymann, F., Reiter, M.: Conventional Workflow Technology for Scientific Simulation, Guide to e-Science, pp. 1–31. Springer (March 2011)
118. Goujon, M., McWilliam, H., Li, W., et al.: A new bioinformatics analysis tools framework at EMBL-EBI. Nucleic Acids Research 38(Web Server issue), W695–W699 (2010)
119. Grady, J., Campbell, H., Faulk, S.R., Weiss, D.M.: Introduction to Synthesis. Technical report, Software Productivity Consortium (June 1990)
120. Groth, P., Gil, Y.: Analyzing the Gap between Workflows and their Natural Language Descriptions. In: IEEE Congress on Services, pp. 299–305. IEEE Computer Society, Los Alamitos (2009)
121. Gubala, T., Herezlak, D., Bubak, M., Malawski, M.: Semantic Composition of Scientific Workflows Based on the Petri Nets Formalism. In: Second IEEE International Conference on e-Science and Grid Computing (e-Science 2006), p. 12. IEEE Computer Society, Washington, DC (2006)
122. Gubała, T., Bubak, M., Malawski, M., Rycerz, K.: Semantic-Based Grid Workflow Composition. In: Wyrzykowski, R., Dongarra, J., Meyer, N., Waśniewski, J. (eds.) PPAM 2005. LNCS, vol. 3911, pp. 651–658. Springer, Heidelberg (2006)
123. Habegger, L., Sboner, A., Gianoulis, T.A., et al.: RSEQtools: a modular framework to analyze RNA-Seq data using compact, anonymized data summaries. Bioinformatics 27(2), 281–283 (2011)
124. Hagemeier, D.: GeneFisher2 - an AJAX based implementation of GeneFisher-P. Bachelor's thesis, University Bielefeld, Faculty of Technology (December 2006)
125. Hahne, F., Huber, W., Gentleman, R., Falcon, S.: Bioconductor Case Studies. Springer (2008)
126. Hartman, A., Riddle, S., McPhillips, T., Ludascher, B., Eisen, J.: Introducing W.A.T.E.R.S.: a Workflow for the Alignment, Taxonomy, and Ecology of Ribosomal Sequences. BMC Bioinformatics 11(1), 317 (2010)
127. Hartmeier, S., Krüger, J., Giegerich, R.: Webservices and Workflows on the Bielefeld Bioinformatics Server: Practices and Problems. In: Proceedings of NETTAB 2007 - A Semantic Web for Bioinformatics, Pisa, Italy (2007)
128. Haubrock, M., Sauer, T., Schwarzer, K., et al.: A Workflow for Retrieving Orthologous Promoters and Implications for Workflow Management Systems. A Case Study. In. In: Proceedings of NETTAB 2007 - A Semantic Web for Bioinformatics, Pisa, Italy (June 2007)
129. Henikoff, S., Henikoff, J.G.: Amino acid substitution matrices from protein blocks. Proceedings of the National Academy of Sciences of the United States of America 89(22), 10915–10919 (1992)
130. Hennessy, M., Milner, R.: Algebraic laws for nondeterminism and concurrency. Journal of the ACM 32, 137–161 (1985)
131. Holmes, I.: Bioinformatics Workflows - BioWiki (January 2006), http://biowiki.org/BioinformaticsWorkflows (last accessed May 5, 2013)

132. Hubbard, T., Barker, D., Birney, E., et al.: The Ensembl genome database project. Nucleic Acids Research 30(1), 38–40 (2002)
133. Huhns, M.N., Singh, M.P.: Service-Oriented Computing: Key Concepts and Principles. IEEE Internet Computing 9, 75–81 (2005)
134. Hull, D., Wolstencroft, K., Stevens, R., et al.: Taverna: a tool for building and running workflows of services. Nucleic Acids Research 34(Web Server), W729–W732 (2006)
135. Irizarry, R.A., Bolstad, B.M., Collin, F., et al.: Summaries of Affymetrix GeneChip probe level data. Nucleic Acids Research 31(4), e15 (2003)
136. Irizarry, R.A., Hobbs, B., Collin, F., et al.: Exploration, normalization, and summaries of high density oligonucleotide array probe level data. Biostatistics 4(2), 249–264 (2003)
137. Irizarry, R.A., Wu, Z., Cawley, S.: affycomp: Graphics Toolbox for Assessment of Affymetrix Expression Measures (2011), http://bioc.ism.ac.jp/2.6/bioc/html/affycomp.html (last accessed May 5, 2013)
138. Ison, J., Kalaš, M., Jonassen, I., et al.: EDAM: an ontology of bioinformatics operations, types of data and identifiers, topics and formats. Bioinformatics (2013)
139. Issarny, V., Steffen, B., Jonsson, B., et al.: CONNECT Challenges: Towards Emergent Connectors for Eternal Networked Systems. In: ICECCS, pp. 154–161. IEEE Computer Society (June 2009)
140. Jordan, D., Evdemon, J., Alves, A., et al.: Web Services Business Process Execution Language Version 2.0 - OASIS Standard (April 2007), http://docs.oasis-open.org/wsbpel/2.0/OS/wsbpel-v2.0-OS.html (last accessed May 5, 2013)
141. Jörges, S., Kubczak, C., Pageau, F., Margaria, T.: Model Driven Design of Reliable Robot Control Programs Using the jABC. In: Proceedings of 4th IEEE International Workshop on Engineering of Autonomic and Autonomous Systems (EASe 2007), pp. 137–148 (2007)
142. Jörges, S.: Construction and Evolution of Code Generators. LNCS, vol. 7747. Springer, Heidelberg (2013)
143. Jörges, S., Lamprecht, A.-L., Margaria, T., Schaefer, I., Steffen, B.: A Constraint-based Variability Modeling Framework. International Journal on Software Tools for Technology Transfer (STTT) 14(5), 511–530 (2012)
144. Jörges, S., Margaria, T., Steffen, B.: Genesys: service-oriented construction of property conform code generators. ISSE 4(4), 361–384 (2008)
145. Kaminuma, E., Kosuge, T., Kodama, Y., et al.: DDBJ progress report. Nucleic Acids Research 39(Database), D22–D27 (2010)
146. Kanehisa, M., Goto, S.: KEGG: Kyoto Encyclopedia of Genes and Genomes. Nucleic Acids Research 28(1), 27–30 (2000)
147. Karlsson, J., Martín-Requena, V., Ríos, J., Trelles, O.: Workflow Composition and Enactment Using jORCA. In: Margaria, T., Steffen, B. (eds.) ISoLA 2010, Part I. LNCS, vol. 6415, pp. 328–339. Springer, Heidelberg (2010)
148. Katayama, T., Arakawa, K., Nakao, M., et al.: The DBCLS BioHackathon: standardization and interoperability for bioinformatics web services and workflows. Journal of Biomedical Semantics 1(8) (2010)
149. Katoen, J.-P.: Labelled Transition Systems. In: Broy, M., Jonsson, B., Katoen, J.-P., Leucker, M., Pretschner, A. (eds.) Model-Based Testing of Reactive Systems. LNCS, vol. 3472, pp. 615–616. Springer, Heidelberg (2005)

150. Katoh, K., Misawa, K., Ichi Kuma, K., Miyata, T.: MAFFT: a novel method for rapid multiple sequence alignment based on fast Fourier transform. Nucleic Acids Research 30(14), 3059–3066 (2002)

151. Kelly, S., Tolvanen, J.-P.: Domain-Specific Modeling: Enabling Full Code Generation. Wiley-IEEE Computer Society Press (2008)

152. Kluge, W.: Reduction, data flow and control flow models of computation. In: Brauer, W., Reisig, W., Rozenberg, G. (eds.) APN 1986. LNCS, vol. 255, pp. 466–498. Springer, Heidelberg (1987)

153. Knoop, J., Steffen, B., Vollmer, J.: Parallelism for free: efficient and optimal bitvector analyses for parallel programs. ACM Transactions on Programming Languages and Systems (TOPLAS) 18(3), 268–299 (1996)

154. Ko, A.J., Abraham, R., Beckwith, L., et al.: The state of the art in end-user software engineering. ACM Computing Surveys 43, 21:1–21:44 (2011)

155. Kona, S., Bansal, A., Blake, M., Gupta, G.: Generalized Semantics-Based Service Composition. In: 2008 IEEE International Conference on Web Services (ICWS 2008), pp. 219–227. IEEE Computer Society (September 2008)

156. Kosakovsky Pond, S., Wadhawan, S., Chiaromonte, F., et al.: Windshield splatter analysis with the Galaxy metagenomic pipeline. Genome Research 19(11), 2144–2153 (2009)

157. Kubczak, C., Jörges, S., Margaria, T., Steffen, B.: eXtreme Model-Driven Design with jABC. In: CTIT Proc. of the Tools and Consultancy Track of the Fifth European Conference on Model-Driven Architecture Foundations and Applications (ECMDA-FA), vol. WP09-12, pp. 78–99 (2009)

158. Kubczak, C., Margaria, T., Fritsch, A., Steffen, B.: Biological LC/MS Preprocessing and Analysis with jABC, jETI and xcms. In: Proceedings of the 2nd International Symposium on Leveraging Applications of Formal Methods, Verification and Validation (ISoLA 2006), Paphos, Cyprus, November 15-19, pp. 308–313. IEEE Computer Society (2006)

159. Kubczak, C., Margaria, T., Kaiser, M., Lemcke, J., Knuth, B.: Abductive Synthesis of the Mediator Scenario with jABC and GEM. In: Semantic Web Services Challenge: Proceedings of the 2008 Workshops, pp. 52–63 (2008)

160. Kubczak, C., Margaria, T., Nagel, R., Steffen, B.: Plug and Play with FMICS-jETI: Beyond Scripting and Coding. ERCIM News (73), 41–42 (2008)

161. Kubczak, C., Margaria, T., Steffen, B.: Mashup Development for Everybody: A Planning-Based Approach. In: Proceedings of the 3rd International SMR2 2009 Workshop on Service Matchmaking and Resource Retrieval in the Semantic Web. CEUR Workshop Proceedings, vol. 525 (October 2009)

162. Kubczak, C., Margaria, T., Steffen, B., Naujokat, S.: Service-oriented Mediation with jETI/jABC: Verification and Export. In: Proceedings of the 2007 IEEE/WIC/ACM International Conference on Web Intelligence and Intelligent Agent Technology, WI-IAT Workshop, pp. 144–147. IEEE Computer Society Press, Silicon Valley (2007)

163. Kupferman, O., Vardi, M.Y.: μ-Calculus Synthesis. In: Nielsen, M., Rovan, B. (eds.) MFCS 2000. LNCS, vol. 1893, pp. 497–507. Springer, Heidelberg (2000)

164. Kutschera, U., Niklas, K.J.: The modern theory of biological evolution: an expanded synthesis. Naturwissenschaften 91(6), 255–276 (2004)

165. Kwon, Y., Shigemoto, Y., Kuwana, Y., Sugawara, H.: Web API for biology with a workflow navigation system. Nucleic Acids Research 37(suppl. 2), W11–W16 (2009)

166. Kühn, K., Greiner, U.: Workflow-Management in der Bioinformatik - Systeme zur Unterstützung in der Bioinformatik/Forschungslaboren. Problemseminar Datenbanken, WS 2001/2002, Abteilung Datenbanken am Institut für Informatik, Universität Leipzig (2002), http://dbs.uni-leipzig.de/html/seminararbeiten/semWS0102/arbeit4/ProblemseminarBioDB_WFM-Bioinf.htm (last accessed May 5, 2013)

167. Labarga, A., Valentin, F., Anderson, M., Lopez, R.: Web services at the European bioinformatics institute. Nucleic Acids Research 35(Web Server issue), W6–W11 (2007)

168. Lamprecht, A.-L.: Intraprocedural data-flow analysis via model checking. Bachelor's thesis, Georg-August-Universität Göttingen (September 2005)

169. Lamprecht, A.-L., Margaria, T., Schaefer, I., Steffen, B.: Comparing Structure-Oriented and Behavior-Oriented Variability Modeling for Workflows. In: Moschitti, A., Scandariato, R. (eds.) EternalS 2011. CCIS, vol. 255, pp. 1–15. Springer, Heidelberg (2012)

170. Lamprecht, A.-L., Margaria, T., Schaefer, I., Steffen, B.: Synthesis-Based Variability Control: Correctness by Construction. In: Beckert, B., Bonsangue, M.M. (eds.) FMCO 2011. LNCS, vol. 7542, pp. 69–88. Springer, Heidelberg (2012)

171. Lamprecht, A.-L., Margaria, T., Steffen, B.: Data-Flow Analysis as Model Checking Within the jABC. In: Mycroft, A., Zeller, A. (eds.) CC 2006. LNCS, vol. 3923, pp. 101–104. Springer, Heidelberg (2006)

172. Lamprecht, A.-L., Margaria, T., Steffen, B.: Seven Variations of an Alignment Workflow - An Illustration of Agile Process Design and Management in Bio-jETI. In: Măndoiu, I., Wang, S.-L., Zelikovsky, A. (eds.) ISBRA 2008. LNCS (LNBI), vol. 4983, pp. 445–456. Springer, Heidelberg (2008)

173. Lamprecht, A.-L., Margaria, T., Steffen, B.: Supporting Process Development in Bio-jETI by Model Checking and Synthesis. In: Semantic Web Applications and Tools for Life Sciences (SWAT4LS 2009). CEUR Workshop Proceedings, vol. 435 (2008)

174. Lamprecht, A.-L., Margaria, T., Steffen, B.: Bio-jETI: a framework for semantics-based service composition. BMC Bioinformatics 10(suppl. 10), S8 (2009)

175. Lamprecht, A.-L., Margaria, T., Steffen, B.: From Bio-jETI Process Models to Native Code. In: 14th IEEE International Conference on Engineering of Complex Computer Systems, ICECCS 2009, Potsdam, Germany, June 2-4, pp. 95–101. IEEE Computer Society (2009)

176. Lamprecht, A.-L., Margaria, T., Steffen, B.: Bioinformatics: Processes and Workflows. In: Laplante, P.A. (ed.) Encyclopedia of Software Engineering, ch. 11, pp. 118–130. Taylor & Francis (November 2010)

177. Lamprecht, A.-L., Margaria, T., Steffen, B., et al.: GeneFisher-P: variations of GeneFisher as processes in Bio-jETI. BMC Bioinformatics 9(suppl. 4), S13 (2008)

178. Lamprecht, A.-L., Naujokat, S., Margaria, T., Steffen, B.: Synthesis-Based Loose Programming. In: Proceedings of the 7th International Conference on the Quality of Information and Communications Technology, QUATIC 2010 (September 2010)

179. Lamprecht, A.-L., Naujokat, S., Margaria, T., Steffen, B.: Semantics-based composition of EMBOSS services. Journal of Biomedical Semantics 2(suppl. 1), S5 (2011)

180. Lamprecht, A.-L., Naujokat, S., Steffen, B., Margaria, T.: Constraint-Guided Workflow Composition Based on the EDAM Ontology. In: Burger, A., Marshall, M.S., Romano, P., Paschke, A., Splendiani, A. (eds.) Proceedings of the 3rd International Workshop on Semantic Web Applications and Tools for Life Sciences (SWAT4LS 2010). CEUR Workshop Proceedings, vol. 698 (December 2010)

181. Lang, D.T.: XML: Tools for parsing and generating XML within R and S-Plus (2009), http://CRAN.R-project.org/package=XML (last accessed May 5, 2013)

182. Larkin, M., Blackshields, G., Brown, N., et al.: Clustal W and Clustal X version 2.0. Bioinformatics 23(21), 2947–2948 (2007)

183. Leymann, F., Roller, D.: Production workflow: concepts and techniques. Prentice Hall PTR, Upper Saddle River (2000)

184. Li, P., Brass, A., Pinney, J., et al.: Taverna workflows for systems biology. In: International Conference on Systems Biology (October 2006)

185. Li, P., Oinn, T., Soiland, S., Kell, D.B.: Automated manipulation of systems biology models using libSBML within Taverna workflows. Bioinformatics 24(2), 287–289 (2008)

186. Lian, C.C., Tang, F., Issac, P., Krishnan, A.: GEL: Grid Execution Languages. Journal of Parallel and Distributed Computing 65, 2005 (2005)

187. Lieberman, H. (ed.): Your Wish is My Command: Programming by Example. Morgan Kaufmann Publishers Inc. (2001)

188. Lieberman, H., Paterno, F., Wulf, V. (eds.): End-User Development. Human-Computer Interaction Series, vol. 9. Springer (2006)

189. Littauer, R., Ram, K., Ludäscher, B., Michener, W., Koskela, R.: Trends in Use of Scientific Workflows: Insights from a Public Repository and Recommendations for Best Practices. In: 7th International Digital Curation Conference (2011)

190. Lord, P., Bechhofer, S., Wilkinson, M.D., Schiltz, G., Gessler, D., Hull, D., Goble, C., Stein, L.: Applying Semantic Web Services to Bioinformatics: Experiences Gained, Lessons Learnt. In: McIlraith, S.A., Plexousakis, D., van Harmelen, F. (eds.) ISWC 2004. LNCS, vol. 3298, pp. 350–364. Springer, Heidelberg (2004)

191. Ludäscher, B., Lin, K., Bowers, S., et al.: Managing Scientific Data: From Data Integration to Scientific Workflows. In: Geoinformatics: Data to Knowledge, Geological Society of America Special Paper 397, pp. 109–129 (2006)

192. Ludäscher, B., Altintas, I., Gupta, A.: Compiling Abstract Scientific Workflows into Web Service Workflows. In: Proceedings of the 15th International Conference on Scientific and Statistical Database Management (SSDBM 2003), pp. 251–254. IEEE Computer Society, Los Alamitos (2003)

193. Ludäscher, B., Weske, M., McPhillips, T., Bowers, S.: Scientific Workflows: Business as Usual? In: Dayal, U., Eder, J., Koehler, J., Reijers, H.A. (eds.) BPM 2009. LNCS, vol. 5701, pp. 31–47. Springer, Heidelberg (2009)

194. Lutz, M.: Ontology-Based Descriptions for Semantic Discovery and Composition of Geoprocessing Services. Geoinformatica 11, 1–36 (2007)

195. Maleki-Dizaji, S., Rolfe, M., Fisher, P., Holcombe, M.: A Systematic Approach to Understanding Bacterial Responses to Oxygen Using Taverna and Webservices. In: 13th International Conference on Biomedical Engineering, vol. 23. Springer, Heidelberg (2009)

196. Manna, Z., Wolper, P.: Synthesis of Communicating Processes from Temporal Logic Specifications. ACM Transactions on Programming Languages and Systems 6(1), 68–93 (1984)

197. Margaria, T.: Service is in the Eyes of the Beholder. IEEE Computer (November 2007)

198. Margaria, T., Bakera, M., Kubczak, C., Naujokat, S., Steffen, B.: Automatic Generation of the SWS-Challenge Mediator with jABC/ABC. In: Petrie, C., Margaria, T., Zaremba, M., Lausen, H. (eds.) Semantic Web Services Challenge. Results from the First Year, pp. 119–138. Springer (2008)

199. Margaria, T., Boßelmann, S., Doedt, M., Floyd, B.D., Steffen, B.: Customer-Oriented Business Process Management: Visions and Obstacles. In: Hinchey, M., Coyle, L. (eds.) Conquering Complexity, pp. 407–429. Springer, London (2012)

200. Margaria, T., Kubczak, C., Njoku, M., Steffen, B.: Model-based Design of Distributed Collaborative Bioinformatics Processes in the jABC. In: Proceedings of the 11th IEEE International Conference on Engineering of Complex Computer Systems (ICECCS 2006), pp. 169–176. IEEE Computer Society, Los Alamitos (2006)

201. Margaria, T., Kubczak, C., Steffen, B.: Bio-jETI: a service integration, design, and provisioning platform for orchestrated bioinformatics processes. BMC Bioinformatics 9(suppl. 4), S12 (2008)

202. Margaria, T., Nagel, R., Steffen, B.: jETI: A Tool for Remote Tool Integration. In: Halbwachs, N., Zuck, L.D. (eds.) TACAS 2005. LNCS, vol. 3440, pp. 557–562. Springer, Heidelberg (2005)

203. Margaria, T., Nagel, R., Steffen, B.: Remote Integration and Coordination of Verification Tools in JETI. In: Proc. of 12th IEEE International Conference on the Engineering of Computer-Based Systems, pp. 431–436. IEEE Computer Society, Los Alamitos (2005)

204. Margaria, T., Steffen, B.: Backtracking-free Design Planning by Automatic Synthesis in METAFrame. In: Astesiano, E. (ed.) ETAPS 1998 and FASE 1998. LNCS, vol. 1382, pp. 188–204. Springer, Heidelberg (1998)

205. Margaria, T., Steffen, B.: Lightweight coarse-grained coordination: a scalable system-level approach. STTT 5(2-3), 107–123 (2004)

206. Margaria, T., Steffen, B.: Service Engineering: Linking Business and IT. Computer 39(10), 45–55 (2006)

207. Margaria, T., Steffen, B.: LTL-Guided Planning: Revisiting Automatic Tool Composition in ETI. In: Proceedings of the 31st IEEE Software Engineering Workshop, pp. 214–226. IEEE Computer Society (2007)

208. Margaria, T., Steffen, B.: Agile IT: Thinking in User-Centric Models. In: Margaria, T., Steffen, B. (eds.) ISoLA 2008. CCIS, vol. 17, pp. 490–502. Springer, Heidelberg (2008)

209. Margaria, T., Steffen, B.: Business Process Modelling in the jABC: The One-Thing-Approach. In: Cardoso, J., van der Aalst, W. (eds.) Handbook of Research on Business Process Modeling. IGI Global (2009)

210. Margaria, T., Steffen, B.: Continuous Model-Driven Engineering. IEEE Computer 42(10), 106–109 (2009)

211. Margaria, T., Steffen, B.: Service-Orientation: Conquering Complexity with XMDD. In: Hinchey, M., Coyle, L. (eds.) Conquering Complexity, pp. 217–236. Springer, London (2012)

212. Margaria, T., Steffen, B., Reitenspieß, M.: Service-Oriented Design: The Roots. In: Benatallah, B., Casati, F., Traverso, P. (eds.) ICSOC 2005. LNCS, vol. 3826, pp. 450–464. Springer, Heidelberg (2005)

213. Margaria, T., Steffen, B., Topnik, C.: Second-Order Value Numbering. Electronic Communications of the EASST (ECEASST) 30 (2010)

214. Martin, D., et al.: Bringing Semantics to Web Services: The OWL-S Approach. In: Cardoso, J., Sheth, A.P. (eds.) SWSWPC 2004. LNCS, vol. 3387, pp. 26–42. Springer, Heidelberg (2005)

215. Martín-Requena, V., Ríos, J., García, M., Ramírez, S., Trelles, O.: jORCA: easily integrating bioinformatics Web Services. Bioinformatics 26(4), 553–559 (2010)

216. May, C.: Entwicklung einer Bibliothek zur service-orientierten Modellierung von Ontologien. Diploma thesis, TU Dortmund (2009)

217. May, P., Ehrlich, H.-C., Steinke, T.: ZIB Structure Prediction Pipeline: Composing a Complex Biological Workflow Through Web Services. In: Nagel, W.E., Walter, W.V., Lehner, W. (eds.) Euro-Par 2006. LNCS, vol. 4128, pp. 1148–1158. Springer, Heidelberg (2006)

218. Mendel, G.: Versuche über Pflanzen-Hybriden. In: Verhandlungen des Naturforschenden Vereins zu Brünn, vol. 4, pp. 3–47 (1866); Separatabdruck aus dem IV. Bande der Verhandlungen des naturforschenden Vereines; mit handschriftlichen Korrekturen Mendels

219. Merten, M., Howar, F., Steffen, B., Cassel, S., Jonsson, B.: Demonstrating Learning of Register Automata. In: Flanagan, C., König, B. (eds.) TACAS 2012. LNCS, vol. 7214, pp. 466–471. Springer, Heidelberg (2012)

220. Merten, M., Steffen, B., Howar, F., Margaria, T.: Next Generation LearnLib. In: Abdulla, P.A., Leino, K.R.M. (eds.) TACAS 2011. LNCS, vol. 6605, pp. 220–223. Springer, Heidelberg (2011)

221. Michener, C.D., Sokal, R.R.: A Quantitative Approach to a Problem in Classification. Evolution 11(2), 130–162 (1957)

222. Mitchell, T.: Web Mapping Illustrated. O'Reilly Media, Inc. (2005)

223. Miyazaki, S., Sugawara, H., Ikeo, K., Gojobori, T., Tateno, Y.: DDBJ in the stream of various biological data. Nucleic Acids Research 32(Database issue), D31–D34 (2004)

224. Monson-Haefel, R. (ed.): 97 Things Every Software Architect Should Know, 1st edn. O'Reilly Media, Inc. (2009)

225. Montali, M., Pesic, M., van der Aalst, W.M.P., et al.: Declarative specification and verification of service choreographiess. TWEB 4(1) (2010)

226. Morgenstern, B.: DIALIGN: multiple DNA and protein sequence alignment at BiBiServ. Nucleic Acids Research 32(suppl. 2), W33–W36 (2004)

227. Myllymäki, T.: Variability management in software product-lines. Technical report 30, Institute of Software Systems, Tampere University of Technology (January 2002)

228. Müller-Olm, M., Schmidt, D.A., Steffen, B.: Model-Checking - A Tutorial Introduction. In: Cortesi, A., Filé, G. (eds.) SAS 1999. LNCS, vol. 1694, pp. 330–354. Springer, Heidelberg (1999)

229. Nagel, R.: Technische Herausforderungen modellgetriebener Beherrschung von Prozesslebenszyklen aus der Fachperspektive von der Anforderungsanalyse zur Realisierung. Dissertation, Technische Universität Dortmund (July 2009)

230. Nanchen, A., Fuhrer, T., Sauer, U.: Determination of metabolic flux ratios from 13C-experiments and gas chromatography-mass spectrometry data: protocol and principles. Methods in Molecular Biology 358, 177–197 (2007)

231. Nau, D., Ilghami, O., Kuter, U., et al.: SHOP2: An HTN planning system. Journal of Artificial Intelligence Research 20, 379–404 (2003)

232. Naujokat, S.: Automatische Generierung von Prozessen im jABC. Diploma thesis, TU Dortmund (September 2009)

233. Naujokat, S., Lamprecht, A.-L., Steffen, B.: Tailoring Process Synthesis to Domain Characteristics. In: Proceedings of the 16th IEEE International Conference on Engineering of Complex Computer Systems, ICECCS (2011)

234. Naujokat, S., Lamprecht, A.-L., Steffen, B.: Loose Programming with PROPHETS. In: de Lara, J., Zisman, A. (eds.) FASE 2012. LNCS, vol. 7212, pp. 94–98. Springer, Heidelberg (2012)

235. Naujokat, S., Neubauer, J., Lamprecht, A.-L., et al.: Plug-Ins: The Enabling Power of Simplicity. In: Garbervetsky, D., Kim, S. (eds.) Special Issue on Tools as Plug-ins, Software: Practice and Experience. John Wiley & Sons, Ltd. (submission; review pending, 2013)

236. Needleman, S.B., Wunsch, C.D.: A general method applicable to the search for similarities in the amino acid sequence of two proteins. Journal of Molecular Biology 48(3), 443–453 (1970)

237. Nelson, B.J.: Remote Procedure Call. Phd thesis, Xerox Palo Alto Research Center (May 1981)

238. Neubauer, J., Margaria, T., Steffen, B.: Design for Verifiability: The OCS Case Study. In: Formal Methods for Industrial Critical Systems: A Survey of Applications, ch. 8, pp. 153–178. Wiley-IEEE Computer Society Press (March 2013)

239. Neubauer, J., Steffen, B.: Second-Order Servification. In: Herzwurm, G., Margaria, T. (eds.) ICSOB 2013. LNBIP, vol. 150, pp. 13–25. Springer, Heidelberg (2013)

240. Nixon, T., Regnier, A., Modi, V., Kemp, D.: Web Services Dynamic Discovery (WS-Discovery) Version 1.1 - OASIS Standard (July 2009), http://docs.oasis-open.org/ws-dd/discovery/1.1/os/wsdd-discovery-1.1-spec-os.pdf

241. Object Management Group (OMG): Common Object Request Broker Architecture (CORBA), http://www.omg.org/spec/CORBA/ (last accessed May 5, 2013)

242. Oinn, T., Greenwood, M., Addis, M., et al.: Taverna: lessons in creating a workflow environment for the life sciences: Research Articles. Concurrency and Computation: Practice and Experience 18(10), 1067–1100 (2006)

243. Oldfield, P.: Domain Modelling (2002), http://www.aptprocess.com/whitepapers/DomainModelling.pdf (last accessed May 5, 2013)

244. Oliver, H., Diallo, G., de Quincey, E., et al.: A user-centred evaluation framework for the Sealife semantic web browsers. BMC Bioinformatics 10(suppl. 10), S14 (2009)

245. Pautasso, C.: A Flexible System for Visual Service Composition. PhD thesis, ETH Zürich (July 2004)

246. Pesic, M., Schonenberg, H., van der Aalst, W.M.P.: DECLARE: Full Support for Loosely-Structured Processes. In: Proceedings of the 11th IEEE International Enterprise Distributed Object Computing Conference (EDOC 2007), pp. 287–300 (2007)

247. Pesic, M., Schonenberg, M.H., Sidorova, N., van der Aalst, W.M.P.: Constraint-based workflow models: Change made easy. In: Meersman, R., Tari, Z. (eds.) OTM 2007, Part I. LNCS, vol. 4803, pp. 77–94. Springer, Heidelberg (2007)

248. Petrie, C., Margaria, T., Lausen, H., Zaremba, M. (eds.): Semantic Web Services Challenge. Results from the First Year. Semantic Web and Beyond, vol. 8. Springer, US (2009)

249. Pettifer, S., Ison, J., Kalas, M., et al.: The EMBRACE web service collection. Nucleic Acids Research (May 2010)

250. Pillai, S., Silventoinen, V., Kallio, K., et al.: SOAP-based services provided by the European Bioinformatics Institute. Nucleic Acids Research 33(Web Server issue), W25–W28 (2005)

251. Pipek, V., Rosson, M.B., de Ruyter, B., Wulf, V. (eds.): IS-EUD 2009. LNCS, vol. 5435. Springer, Heidelberg (2009)

252. Pnueli, A., Rosner, R.: On the synthesis of a reactive module. In: Annual Symposium on Principles of Programming Languages (1989)

253. Pohl, K., Böckle, G., van der Linden, F.J.: Software Product Line Engineering: Foundations, Principles and Techniques. Springer-Verlag New York, Inc., Secaucus (2005)

254. Polanski, A., Kimmel, M.: Bioinformatics, 1st edn. Springer, Heidelberg (2007)

255. Poole, D., Mackworth, A.: Artificial Intelligence: Foundations of Computational Agents. Cambridge University Press, New York (2010)

256. Potter, S., Aitken, S.: A Semantic Service Environment: A Case Study in Bioinformatics. In: Gómez-Pérez, A., Euzenat, J. (eds.) ESWC 2005. LNCS, vol. 3532, pp. 694–709. Springer, Heidelberg (2005)

257. Qin, J., Fahringer, T.: A novel domain oriented approach for scientific grid workflow composition. In: Proceedings of the 2008 ACM/IEEE Conference on Supercomputing, SC 2008, pp. 21:1–21:12. IEEE Press, Piscataway (2008)

258. Qin, J., Fahringer, T.: Scientific Workflows - Programming, Optimization, and Synthesis with ASKALON and AWDL. Springer, Heidelberg (2012)

259. Quek, L., Wittmann, C., Nielsen, L., Kromer, J.: OpenFLUX: efficient modelling software for 13C-based metabolic flux analysis. Microbial Cell Factories 8(25) (2009)

260. Raffelt, H., Steffen, B., Berg, T., Margaria, T.: LearnLib: a framework for extrapolating behavioral models. International Journal on Software Tools for Technology Transfer (STTT) 11(5), 393–407 (2009)

261. Ramirez, S., Karlsson, J., Trelles, O.: MAPI: towards the integrated exploitation of bioinformatics Web Services. BMC Bioinformatics 12(1), 419 (2011)

262. Randriamparany, H., Ibrahim, B.: Seamless Integration of Control Flow and Data Flow in a Visual Language. In: Proceedings of the ACS/IEEE International Conference on Computer Systems and Applications, pp. 428–434. IEEE Computer Society, Washington, DC (2001)

263. Reich, M., Liefeld, T., Gould, J., et al.: GenePattern 2.0. Nature Genetics 38(5), 500–501 (2006)

264. Rice, P., Longden, I., Bleasby, A.: EMBOSS: the European Molecular Biology Open Software Suite. Trends in Genetics 16(6), 276–277 (2000)

265. Rios, J., Karlsson, J., Trelles, O.: Magallanes: a web services discovery and automatic workflow composition tool. BMC Bioinformatics 10(1), 334 (2009)

266. Russel, S., Norvig, P.: Artificial Intelligence: A Modern Approach, 3rd edn. Prentice Hall (December 2009)

267. Sahoo, S.S., Sheth, A., Henson, C.: Semantic Provenance for eScience: Managing the Deluge of Scientific Data. IEEE Internet Computing 12(4), 46–54 (2008)
268. Saitou, N., Nei, M.: The neighbor-joining method: a new method for reconstructing phylogenetic trees. Molecular Biology and Evolution 4(4), 406–425 (1987)
269. Sauer, U., Lasko, D.R., Fiaux, J., et al.: Metabolic flux ratio analysis of genetic and environmental modulations of Escherichia coli central carbon metabolism. Journal of Bacteriology 181(21), 6679–6688 (1999)
270. Sayers, E.W., Barrett, T., Benson, D.A., et al.: Database resources of the National Center for Biotechnology Information. Nucleic Acids Research 39(Database issue), D38–D51 (2011)
271. Schaefer, I., Lamprecht, A.-L., Margaria, T.: Constraint-oriented Variability Modeling. In: Rash, J., Rouff, C. (eds.) 34th Annual IEEE Software Engineering Workshop (SEW-34), pp. 77–83. IEEE CS Press (June 2011)
272. Schaefer, I., Rabiser, R., Clarke, D., et al.: Software Diversity – State of the Art and Perspectives. International Journal on Software Tools for Technology Transfer (STTT) 14, 477–495 (2012)
273. Schmidt, D.A., Steffen, B.: Program Analysis *as* Model Checking of Abstract Interpretations. In: Levi, G. (ed.) SAS 1998. LNCS, vol. 1503, pp. 351–380. Springer, Heidelberg (1998)
274. Schmidt, D.C.: Guest Editor's Introduction: Model-Driven Engineering. IEEE Computer 39(2), 25–31 (2006)
275. Schmidt, K., Carlsen, M., Nielsen, J., Villadsen, J.: Modeling isotopomer distributions in biochemical networks using isotopomer mapping matrices. Biotechnology and Bioengineering 55(6), 831–840 (1997)
276. Schreiber, G., Dean, M.: OWL Web Ontology Language Reference. W3C Recommendation (2004), http://www.w3.org/TR/owl-ref/ (last accessed May 5, 2013)
277. Schroeder, M., Burger, A., Kostkova, P., et al.: From a Service-based eScience Infrastructure to a Semantic Web for the Life Sciences: The Sealife Project. In: Proceedings of the Workshop on Network Tools and Applications in Biology, NETTAB 2006 (2006)
278. Seidl, H., Steffen, B.: Constraint-based inter-procedural analysis of parallel programs. Nordic J. of Computing 7(4), 375–400 (2000)
279. Selzer, P., Marhöfer, R., Rohwer, A.: Applied Bioinformatics: An Introduction, 1st edn. Springer (February 2008)
280. Shannon, P., Markiel, A., Ozier, O., et al.: Cytoscape: a software environment for integrated models of biomolecular interaction networks. Genome Research 13(11), 2498–2504 (2003)
281. Shields, M.: Control- Versus Data-Driven Workflows. In: Taylor, I.J., Deelman, E., Gannon, D.B., Shields, M. (eds.) Workflows for e-Science, pp. 167–173. Springer, London (2007)
282. Shrager, J., Waldinger, R., Stickel, M., Massar, J.P.: Deductive biocomputing. PLoS ONE 2(4), e339 (2007)
283. Simon, S.: Agiles Prozessmanagement: Konzeptuelle Entwicklung und praktische Evaluation auf Basis des jABC Frameworks. Diploma thesis, Technische Universität Dortmund (2009)
284. Sirin, E., Parsia, B., Wu, D., Hendler, J.A., Nau, D.S.: HTN planning for Web Service composition using SHOP2. Journal of Web Semantics 1(4), 377–396 (2004)

285. Smith, B., Ashburner, M., Rosse, C., et al.: The OBO Foundry: coordinated evolution of ontologies to support biomedical data integration. Nature Biotechnology 25(11), 1251–1255 (2007)

286. Smith, C.A.: annaffy: Annotation tools for Affymetrix biological metadata (2008), http://bioc.ism.ac.jp/2.5/bioc/html/annaffy.html (last accessed May 5, 2013)

287. Smith, D.C., Cypher, A., Tesler, L.: Programming by example: novice programming comes of age. Communications of the ACM 43, 75–81 (2000)

288. Smith, T.F., Waterman, M.S.: Identification of common molecular subsequences. Journal of Molecular Biology 147(1), 195–197 (1981)

289. Smolka, S., Steffen, B.: Priority as extremal probability. In: Baeten, J.C.M., Klop, J.W. (eds.) CONCUR 1990. LNCS, vol. 458, pp. 456–466. Springer, Heidelberg (1990)

290. Smyth, G.K.: Limma: linear models for microarray data. In: Gentleman, R., Carey, V., Dudoit, S., Irizarry, W.H.R. (eds.) Bioinformatics and Computational Biology Solutions using R and Bioconductor, pp. 397–420. Springer, New York (2005)

291. Specht, M., Kuhlgert, S., Fufezan, C., Hippler, M.: Proteomics to go: Proteomatic enables the user-friendly creation of versatile MS/MS data evaluation workflows. Bioinformatics (February 2011)

292. Steffen, B.: Characteristic Formulae. In: Ausiello, G., Dezani-Ciancaglini, M., Ronchi Della Rocca, S. (eds.) ICALP 1989. LNCS, vol. 372, pp. 723–732. Springer, Heidelberg (1989)

293. Steffen, B.: Data Flow Analysis as Model Checking. In: Ito, T., Meyer, A.R. (eds.) TACS 1991. LNCS, vol. 526, pp. 346–364. Springer, Heidelberg (1991)

294. Steffen, B.: Generating data flow analysis algorithms from modal specifications. Selected Papers of the Conference on Theoretical Aspects of Computer Software, pp. 115–139. Elsevier Science Publishers B. V., Sendai (1993)

295. Steffen, B.: Generating Data Flow Analysis Algorithms from Modal Specifications. Science of Computer Programming 21(2), 115–139 (1993)

296. Steffen, B.: Property-oriented expansion. In: Cousot, R., Schmidt, D.A. (eds.) SAS 1996. LNCS, vol. 1145, pp. 22–41. Springer, Heidelberg (1996)

297. Steffen, B.: Method for incremental synthesis of a discrete technical system (1998)

298. Steffen, B., Claßen, A., Klein, M., Knoop, J., Margaria, T.: The Fixpoint-Analysis Machine. In: Lee, I., Smolka, S.A. (eds.) CONCUR 1995. LNCS, vol. 962, pp. 72–87. Springer, Heidelberg (1995)

299. Steffen, B., Ingólfsdóttir, A.: Characteristic Formulae for Processes with Divergence. Information and Computation 110(1), 149–163 (1994)

300. Steffen, B., Knoop, J., Rüthing, O.: The Value Flow Graph: A Program Representation for Optimal Program Transformations. In: Jones, N.D. (ed.) ESOP 1990. LNCS, vol. 432, pp. 389–405. Springer, Heidelberg (1990)

301. Steffen, B., Margaria, T., Braun, V.: The Electronic Tool Integration platform: concepts and design. International Journal on Software Tools for Technology Transfer (STTT) 1(1-2), 9–30 (1997)

302. Steffen, B., Margaria, T., Braun, V., Kalt, N.: Hierarchical Service Definition. Annual Review of Communications of the ACM 51, 847–856 (1997)

303. Steffen, B., Margaria, T., Claßen, A., Braun, V.: Incremental Formalization: A Key to Industrial Success. Software - Concepts and Tools 17(2), 78–95 (1996)

304. Steffen, B., Margaria, T., Claßen, A., Braun, V., Reitenspieß, M.: An Environment for the Creation of Intelligent Network Services. In: Intelligent Networks: IN/AIN Technologies, Operations, Services and Applications - A Comprehensive Report, pp. 287–300. IEC: International Engineering Consortium (1996)

305. Steffen, B., Margaria, T., Freitag, B.: Module Configuration by Minimal Model Construction. Technical report, Fakultät für Mathematik und Informatik, Universität Passau (1993)

306. Steffen, B., Margaria, T., Nagel, R., Jörges, S., Kubczak, C.: Model-Driven Development with the jABC. In: Bin, E., Ziv, A., Ur, S. (eds.) HVC 2006. LNCS, vol. 4383, pp. 92–108. Springer, Heidelberg (2007)

307. Steffen, B., Margaria, T., von der Beeck, M.: Automatic synthesis of linear process models from temporal constraints: An incremental approach. In: ACM/SIGPLAN International Workshop on Automated Analysis of Software, AAS 1997 (1997)

308. Steiner, L., Stadler, P.F., Cysouw, M.: A Pipeline for Computational Historical Linguistics. Language Dynamics and Change 1(1), 89–127 (2011)

309. Stevens, R.D., Robinson, A.J., Goble, C.A.: myGrid: personalised bioinformatics on the information grid. Bioinformatics 19(suppl. 1), i302–i304 (2003)

310. Still, M.: The Definitive Guide to ImageMagick. Apress, Berkely (2005)

311. Stojanovic, Z., Dahanayake, A. (eds.): Service-oriented Software System Engineering Challenges and Practices. IGI Publishing, Hershey (2005)

312. Stoye, J.: Multiple sequence alignment with the Divide-and-Conquer method. Gene 211(2), GC45–GC56 (1998)

313. Sutherland, K., McLeod, K., Ferguson, G., Burger, A.: Knowledge-driven enhancements for task composition in bioinformatics. BMC Bioinformatics 10(suppl. 10), S12 (2009)

314. Tan, W., Missier, P., Madduri, R., Foster, I.: Building Scientific Workflow with Taverna and BPEL: A Comparative Study in caGrid. In: Feuerlicht, G., Lamersdorf, W. (eds.) ICSOC 2008. LNCS, vol. 5472, pp. 118–129. Springer, Heidelberg (2009)

315. Tang, F., Chua, C.L., Ho, L., et al.: Wildfire: distributed, Grid-enabled workflow construction and execution. BMC Bioinformatics 6(1), 69 (2005)

316. Tang, W., Selwood, J.: Connecting our world: GIS Web services. ESRI Press (2003)

317. Taylor, I.: Workflows for E-Science: Scientific Workflows for Grids. Springer (2007)

318. Taylor, I., Shields, M., Wang, I., Harrison, A.: The Triana Workflow Environment: Architecture and Applications. In: Workflows for e-Science, ch. 20, pp. 320–339. Springer, New York (2007)

319. Tessier, P., Servat, D., Gérard, S.: Variability Management on Behavioral Models. In: VaMoS 2008, pp. 121–130 (2008)

320. Thompson, J.D., Higgins, D.G., Gibson, T.J.: CLUSTAL W: improving the sensitivity of progressive multiple sequence alignment through sequence weighting, position-specific gap penalties and weight matrix choice. Nucleic Acids Research 22(22), 4673–4680 (1994)

321. Treleaven, P.C., Hopkins, R.P., Rautenbach, P.W.: Combining data flow and control flow computing. The Computer Journal 25(2), 207–217 (1982)

322. U.S. National Center of Biotechnology Information (NCBI): PubMed, http://www.ncbi.nlm.nih.gov/pubmed/ (last accessed May 5, 2013)

323. Valipour, M.H., Amirzafari, B., Maleki, K.N., Daneshpour, N.: A brief survey of software architecture concepts and service oriented architecture. In: 2nd IEEE International Conference on Computer Science and Information Technology, ICCSIT 2009, pp. 34–38 (August 2009)

324. Valmari, A.: The State Explosion Problem. In: Reisig, W., Rozenberg, G. (eds.) APN 1998. LNCS, vol. 1491, pp. 429–528. Springer, Heidelberg (1998)

325. van der Aalst, W.M.P., Pesic, M.: DecSerFlow: Towards a Truly Declarative Service Flow Language. In: Bravetti, M., Núñez, M., Zavattaro, G. (eds.) WS-FM 2006. LNCS, vol. 4184, pp. 1–23. Springer, Heidelberg (2006)

326. van der Aalst, W.M.P., ter Hofstede, A.H.M., Kiepuszewski, B., Barros, A.P.: Workflow Patterns. Distributed and Parallel Databases 14(1), 5–51 (2003)

327. van der Aalst, W.M.P., van Hee, K.: Workflow Management: Models, Methods, and Systems. MIT Press (2002)

328. van der Aalst, W.M.P., Weijters, A.J.M.M., Maruster, L.: Workflow Mining: Discovering Process Models from Event Logs. IEEE Transactions on Knowledge and Data Engineering 16(9), 1128–1142 (2004)

329. Van der Linden, F., Schmid, K., Rommes, E.: Software product lines in action: the best industrial practice in product line engineering. Springer, New York (2007)

330. van Gurp, J., Bosch, J. (eds.): Software Variability Management (2003)

331. Vandervalk, B.P., McCarthy, E.L., Wilkinson, M.D.: SHARE: A Semantic Web Query Engine for Bioinformatics. In: Gómez-Pérez, A., Yu, Y., Ding, Y. (eds.) ASWC 2009. LNCS, vol. 5926, pp. 367–369. Springer, Heidelberg (2009)

332. Vigder, M.R., Vinson, N.G., Singer, J., Stewart, D.A., Mews, K.: Supporting Scientists' Everyday Work: Automating Scientific Workflows. IEEE Software 25(4), 52–58 (2008)

333. Visser, U., Stuckenschmidt, H., Schuster, G., Vögele, T.: Ontologies for geographic information processing. Computers & Geosciences 28, 103–117 (2002)

334. von Musil, C.: jR - Integration von Statistikfunktionalität in eine Prozessmanagementumgebung. Diplomarbeit, Technische Universität Dortmund (September 2009)

335. Voorhoeve, M., van der Aalst, W.M.P.: Ad-hoc workflow: problems and solutions. In: Proceedings of the 8th International Workshop on Database and Expert Systems Applications (DEXA 1997), pp. 36–40. IEEE Computer Society, Washington, DC (1997)

336. Wassink, I., Rauwerda, H., van der Vet, P., Breit, T., Nijholt, A.: E-BioFlow: Different Perspectives on Scientific Workflows. In: Elloumi, M., Küng, J., Linial, M., Murphy, R.F., Schneider, K., Toma, C. (eds.) BIRD 2008. CCIS, vol. 13, pp. 243–257. Springer, Heidelberg (2008)

337. Weijters, A.J.M.M., van der Aalst, W.M.P.: Process Mining: Discovering Workflow Models from Event-Based Data. In: Proceedings of the 13th Belgium-Netherlands Conference on Artificial Intelligence (BNAIC 2001), pp. 283–290 (2001)

338. White, T.J., Arnheim, N., Erlich, H.A.: The polymerase chain reaction. Trends in Genetics: TIG 5(6), 185–189 (1989)

339. Wiechert, W.: 13C metabolic flux analysis. Metabolic Engineering 3(3), 195–206 (2001)

340. Wikipedia: Bioinformatics workflow management systems — Wikipedia, The Free Encyclopedia, `http://en.wikipedia.org/wiki/Bioinformatics_workflow_management_systems` (last accessed May 5, 2013)

341. Wilkinson, M.D., Links, M.: BioMOBY: an open source biological web services proposal. Briefings in Bioinformatics 3(4), 331–341 (2002)
342. Wilkinson, M.D., Vandervalk, B., McCarthy, L.: SADI Semantic Web Services - 'cause you can't always GET what you want! In: Proceedings of the IEEE Services Computing Conference, APSCC 2009, Singapore, December 7-11, pp. 13–18. IEEE Asia-Pacific (2009)
343. Wilkinson, M.D., Vandervalk, B., McCarthy, L.: The Semantic Automated Discovery and Integration (SADI) Web service Design-Pattern, API and Reference Implementation. Journal of Biomedical Semantics 2(1), 8 (2011)
344. Withers, D., Kawas, E., McCarthy, L., Vandervalk, B., Wilkinson, M.: Semantically-guided workflow construction in Taverna: The SADI and BioMoby plug-ins. In: Margaria, T., Steffen, B. (eds.) ISoLA 2010, Part I. LNCS, vol. 6415, pp. 301–312. Springer, Heidelberg (2010)
345. Wittmann, C.: Fluxome analysis using GC-MS. Microbial Cell Factories 6, 6 (2007)
346. Woese, C.R., Kandler, O., Wheelis, M.L.: Towards a natural system of organisms: proposal for the domains Archaea, Bacteria, and Eucarya. Proceedings of the National Academy of Sciences of the United States of America 87(12), 4576–4579 (1990)
347. Wolstencroft, K., Alper, P., Hull, D., et al.: The (my)Grid ontology: bioinformatics service discovery. International Journal of Bioinformatics Research and Applications 3(3), 303–325 (2007)
348. Wood, I., Vandervalk, B., McCarthy, L., Wilkinson, M.D.: OWL-DL Domain-Models as Abstract Workflows. In: Margaria, T., Steffen, B. (eds.) ISoLA 2012, Part II. LNCS, vol. 7610, pp. 56–66. Springer, Heidelberg (2012)
349. Wu, J., Liu, X., Gao, G., et al.: WebLab: a bioinformatics platform. EMBnet.news 12(2), 13–14 (2006)
350. Wu, X., Li, X.A.: AffyExpress: Affymetrix Quality Assessment and Analysis Tools (2008), http://bioc.ism.ac.jp/2.6/bioc/html/AffyExpress.html (last accessed May 5, 2013)
351. Wu, Z., Irizarry, R., MacDonal, J., Gentry, J.: gcrma: Background Adjustment Using Sequence Information (2007), http://bioc.ism.ac.jp/2.5/bioc/html/gcrma.html (last accessed May 5, 2013)
352. Yue, P., Di, L., Yang, W., Yu, G., Zhao, P.: Semantics-based automatic composition of geospatial Web service chains. Computers & Geosciences 33, 649–665 (2007)
353. Zamboni, N., Fischer, E., Sauer, U.: FiatFlux – a software for metabolic flux analysis from 13C-glucose experiments. BMC Bioinformatics 6, 209 (2005)
354. Zeng, R., He, X., van der Aalst, W.M.P.: A Method to Mine Workflows from Provenance for Assisting Scientific Workflow Composition. In: 7th IEEE World Congress on Services (SERVICES 2011), pp. 169–175. IEEE Computer Society, Los Alamitos (2011)
355. Zhao, Y., Raicu, I., Foster, I.: Scientific Workflow Systems for 21st Century, New Bottle or New Wine? In: Proceedings of the 2008 IEEE Congress on Services - Part I, pp. 467–471. IEEE Computer Society (2008)
356. Zhao, Z., Belloum, A., Wibisono, A., et al.: Scientific workflow management: between generality and applicability. In: Proceedings of the Fifth International Conference on Quality Software, QSIC 2005, pp. 357–364. IEEE Computer Society, Washington, DC (2005)